Introduction to Autonomous Driving

Weisong Shi • Yuankai He

Introduction to Autonomous Driving

Springer

Weisong Shi
Computer and Information Sciences
University of Delaware
Newark, DE, USA

Yuankai He
University of Delaware
Newark, DE, USA

ISBN 978-3-031-99484-5 ISBN 978-3-031-99485-2 (eBook)
https://doi.org/10.1007/978-3-031-99485-2

Cover illustration: Courtesy of the CAR Lab.

This Springer imprint is published by the registered company Springer Nature Switzerland AG
The registered company address is: Gewerbestrasse 11, 6330 Cham, Switzerland

Preface

The field of autonomous driving stands at the intersection of artificial intelligence, robotics, embedded systems, and transportation engineering. Over the past two decades, advances in perception algorithms, sensor integration, real-time control, and vehicle-to-everything (V2X) communication have moved autonomous vehicles (AVs) from speculative prototypes to road-tested systems with broad industrial and societal implications.

This book was conceived as both a comprehensive introduction and a hands-on guide for students, researchers, and practitioners entering the AV ecosystem. It reflects the collaborative spirit of a multidisciplinary field, drawing on the collective expertise of contributors with backgrounds in computing systems, control theory, embedded development, and ethical governance. Each chapter is constructed to bridge conceptual knowledge with real-world application—through the use of simulation environments like BlueICE, real-world datasets, and modular tools such as Autoware.Universe.

A unique aspect of this book is its emphasis on experiential learning. From simulating sensor fusion in BlueICE to implementing planning strategies and security protocols, learners are encouraged to engage directly with the complexities and ambiguities of autonomous vehicle development. The integration of reflection exercises on ethics and society underlines our belief that technical fluency must be accompanied by responsible innovation.

The organization of the text—from foundational chapters on perception and localization to advanced topics such as end-to-end systems and the AV industry landscape—mirrors the layered architecture of an actual autonomous vehicle. Our goal is to equip readers not only with technical skill but with the conceptual clarity and critical perspective necessary for leadership in this evolving domain.

We are especially grateful to Yuxin Wang, Lichen Xia, Zhaofeng Tian, and Ren Zhong for their thoughtful contributions to this book. Their technical insight, instructional clarity, and domain expertise enriched the depth and scope of the work.

This book is the result of sustained effort from collaborators from both academia and industry, and former and current students from the CAR lab at Wayne State University and the University of Delaware, whose contributions and feedback shaped its structure and pedagogical intent. We hope it serves as a catalyst for exploration, experimentation, and ethical practice in the autonomous driving community.

Newark, DE, USA Weisong Shi
May 2025 Yuankai He

Declarations

Competing Interests The authors have no competing interests to declare that are relevant to the content of this manuscript.

Contents

Chapter 1
Introduction to Autonomous Driving

Autonomous vehicles (AVs) are not just a technological evolution; they are a paradigm shift that will redefine the very fabric of modern society, revolutionizing transportation, urban planning, and the global economy. Imagine a world where road fatalities are nearly eliminated, traffic congestion is a thing of the past, and mobility is a seamless, efficient experience for everyone—regardless of physical ability, age, or socio-economic background. Envision smart cities where autonomous taxis replace private car ownership, where real-time traffic optimization minimizes delays, and where freight logistics are automated to maximize efficiency and reduce costs. AVs leverage cutting-edge artificial intelligence (AI), sensor fusion, real-time data analytics, and machine learning to create sophisticated, adaptive systems capable of perceiving, interpreting, and responding to their surroundings in real-time. These intelligent systems analyze a constant stream of data from cameras, LiDAR, radar, and other sensors, making instantaneous decisions that mimic and, in many cases, surpass human capabilities.

These vehicles are more than just sophisticated machines; they are catalysts for a fundamental transformation in how we design cities, conduct business, and interact with technology. The ripple effects of AV adoption extend far beyond individual transportation—they influence everything from urban infrastructure and traffic management to job markets and sustainability policies. AV technology has the potential to redefine everything from personal commuting and freight logistics to emergency response systems and public transportation. Governments and industries worldwide are investing billions into this transformative technology, aiming to unlock a future where vehicles operate seamlessly, communicating with each other and the surrounding environment to create an efficient and safer road network.

This chapter provides an in-depth exploration of the fundamental principles of AVs, detailing their definitions, classification levels, historical breakthroughs, and the multifaceted ways in which they are poised to reshape industries, economies, and daily life. By understanding these foundational concepts, students and researchers can better appreciate the complexities involved in developing, implementing,

W. Shi, Y. He, *Introduction to Autonomous Driving*,
https://doi.org/10.1007/978-3-031-99485-2_1

and regulating autonomous vehicle technology. Additionally, we will examine the formidable technical challenges that engineers, policymakers, and researchers must address to make full autonomy a reality across diverse and unpredictable environments.

1.1 Definitions of Autonomous Vehicles and SAE Levels of Autonomy

The foundational understanding of autonomous vehicles is well-supported by a range of surveys and reviews that categorize the technological, regulatory, and ethical landscape. Works such as those by Khan et al. [6], Rosenzweig and Bartl [12] offer comprehensive overviews of the autonomous driving pipeline and the gap between current capabilities and the vision of full autonomy. These studies examine perception, planning, control, and system integration in great detail, forming a conceptual base for further research.

Technical innovations in machine learning and AI are central to autonomous driving, as outlined in papers by Huang and Chen [5], [2], and Berkenkamp et al. [1]. These contributions focus on deep learning models, cognitive systems, and explainable AI, emphasizing their role in real-time perception and planning. Additionally, Yurtsever et al. [18] and Omeiza et al. [11] elaborate on common practices and the growing need for transparency in AI decision-making, a theme also found in Dong et al.'s vision for collaborative driving [4].

Autonomous vehicles are defined as intelligent systems that integrate advanced sensors, artificial intelligence, and decision-making algorithms to perceive their surroundings, process environmental data, and autonomously navigate roads with minimal or no human intervention. These vehicles rely on a combination of real-time sensor fusion, deep learning models, and predictive analytics to interpret complex traffic scenarios, respond dynamically to changing conditions, and ensure safe navigation. The automation spectrum encompasses a range of capabilities, from basic driver assistance features—such as lane-keeping and adaptive cruise control—to fully autonomous vehicles that can operate seamlessly in any environment without human oversight. As AV technology continues to evolve, researchers and engineers are striving to enhance their ability to handle unpredictable real-world situations, including adverse weather conditions, emergency scenarios, and interactions with human drivers.

To standardize the classification of automation levels, the Society of Automotive Engineers (SAE) introduced a six-level framework [13]. These levels are as follows:

- Level 0: No automation; the driver has full control.
- Level 1: Driver assistance; features like adaptive cruise control assist but do not control the vehicle entirely.

- Level 2: Partial automation; systems can manage both speed and steering but require driver supervision.
- Level 3: Conditional automation; the vehicle can operate autonomously in specific environments but requires human intervention upon request.
- Level 4: High automation; AVs can handle all driving tasks within defined operational conditions, such as urban centers or highways.
- Level 5: Full automation; AVs operate independently in all environments and conditions without human intervention (Figs. 1.1 and 1.2).

Fig. 1.1 A L4 Autonomous Vehicle from the CAR Lab at the University of Delaware

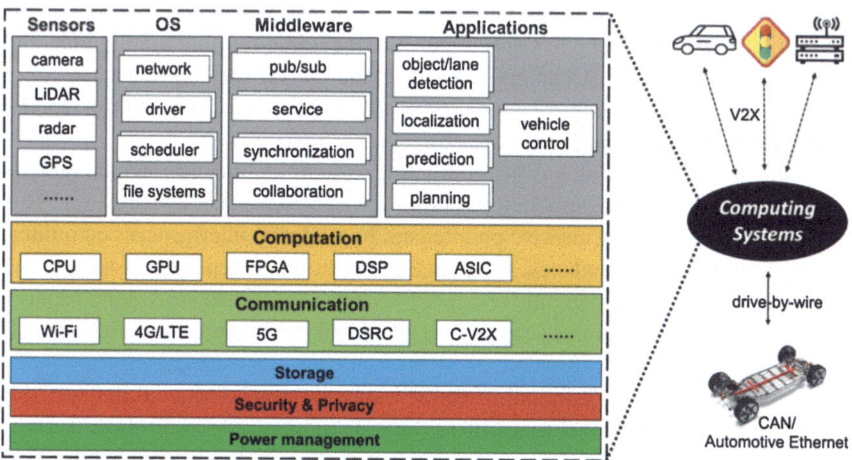

Fig. 1.2 The key components of an autonomous vehicle[8]

1.2 Key Components

Autonomous driving systems are composed of a series of interdependent subsystems that must operate in real time under dynamic, safety-critical conditions. From sensing to actuation, these systems collectively form a computational pipeline that mirrors the perception-decision-action loop observed in biological systems. This section provides an overview of these key components and sets the stage for their detailed treatment in later chapters.

The first essential component is perception, which enables the vehicle to interpret its environment. This includes recognizing objects, estimating their positions and velocities, and understanding the broader scene context. As detailed in Chap. 5, perception relies on high-bandwidth sensor data—such as LiDAR, radar, and cameras—and involves sophisticated algorithms for object detection, semantic segmentation, and anomaly recognition.

Next, localization plays a critical role by continuously estimating the vehicle's position and orientation within a known or constructed map. Modern localization algorithms integrate GPS, LiDAR scans, visual odometry, and inertial measurements to achieve centimeter-level accuracy. Chapter 6 explores various methods including Kalman filtering and SLAM (Simultaneous Localization and Mapping), highlighting the trade-offs in accuracy, robustness, and computational cost.

Prediction extends perception by forecasting the future states of dynamic agents in the environment, such as vehicles and pedestrians. These predictions inform the planning subsystem, which computes feasible and optimal trajectories that adhere to safety, comfort, and efficiency constraints. Planning is examined in Chap. 7, which also introduces the decision-making frameworks required to handle uncertainty and multi-agent interaction, such as Markov Decision Processes and reinforcement learning.

Once a trajectory is planned, control systems ensure it is followed precisely. This includes managing the vehicle's speed and steering angle through longitudinal and lateral control strategies. Chapters 8 and 9 elaborate on the importance of low-latency, high-reliability control loops, typically implemented via PID or Model Predictive Control schemes.

Underlying these modules is the computing system architecture, which orchestrates the data flow and computational tasks across heterogeneous hardware. This architecture must support massive parallelism, low-latency inter-process communication, and fail-safe redundancy. Chapter 9 provides an in-depth discussion of the middleware, hardware accelerators (e.g., GPUs and TPUs), and software interfaces that enable the coordination of perception, planning, and control under strict real-time constraints.

Finally, recent advances in end-to-end learning systems, discussed in Chap. 11, propose a unified model that maps raw sensor data directly to control commands. While promising in terms of reducing system complexity, these models raise new challenges regarding interpretability, robustness, and generalization across diverse driving environments.

1.3 Historical Milestones in Autonomous Vehicle Development

The development of autonomous vehicles has been an extraordinary journey spanning nearly a century, characterized by groundbreaking innovations, interdisciplinary collaboration, and a relentless pursuit of automation. The concept of self-driving machines, once confined to science fiction, has gradually transformed into a tangible reality through iterative advancements in artificial intelligence, robotics, and sensor technologies. From early experiments with guided systems to the modern integration of machine learning and cloud computing, each milestone in AV development has played a crucial role in shaping the trajectory of this transformative field.

The history of autonomous vehicles is not merely a timeline of technological progress; it is a story of human ingenuity, perseverance, and the relentless pursuit of efficiency and safety. In the early twentieth century, automation was little more than an ambitious dream, yet visionaries imagined a future where transportation would be seamless, efficient, and free of human error. As computing power grew and the field of artificial intelligence matured, the once-distant dream of fully autonomous vehicles began to take shape. Over the decades, a combination of academic research, corporate investment, and government-funded initiatives accelerated the pace of AV development, ushering in an era where self-driving vehicles are no longer a fantasy but an impending reality.

Beyond mere transportation, AVs are at the heart of a larger movement toward smart cities, intelligent infrastructure, and interconnected systems. Advances in connectivity, including Vehicle-to-Vehicle (V2V) and Vehicle-to-Infrastructure (V2I) communication, are paving the way for unprecedented levels of efficiency in urban mobility. Meanwhile, ethical considerations, regulatory frameworks, and cybersecurity measures are being actively developed to ensure that AVs contribute positively to society rather than introduce new risks.

More recently, infrastructure and computational architecture receive attention in studies such as those by Liu et al. [9], where edge computing and system scalability are explored. These works argue for modular, latency-aware frameworks capable of handling data in dynamic environments. Sensor fusion, a critical enabler of robust perception, is reviewed by Trypuz et al. [14], who compare probabilistic and learning-based methods.

Papers by Bimbraw [1] and Chen et al. [2] trace the historical trajectory of autonomous vehicle development, offering insights into the evolution of system components, algorithms, and testing frameworks. These historical accounts contextualize how AV systems have matured over the decades.

Below is an expanded timeline detailing the pivotal breakthroughs, visionary projects, and technological leaps that have contributed to the evolution of autonomous vehicles, ultimately paving the way for a future where intelligent transportation systems redefine mobility as we know it.

1.3.1 1939: General Motors' Futurama Concept [17]

General Motors introduced its "Futurama" exhibit at the 1939 New York World's Fair, showcasing a bold vision of an interconnected transportation network where vehicles navigated seamlessly on automated highways controlled by radio signals embedded in roadways. This futuristic depiction not only captured the public's imagination but also set the stage for future research into infrastructure-assisted automation. The exhibit illustrated a world where traffic congestion was eliminated through centralized control systems, vehicles could move efficiently in synchronized flows, and urban environments were designed with automation in mind.

A key aspect of the Futurama exhibit was its emphasis on a highly structured transportation ecosystem, where human error was virtually eliminated by integrating automation into the road infrastructure itself. This concept foresaw the development of automated traffic control centers that could dynamically adjust vehicle movement patterns in real time, much like modern-day adaptive traffic signal systems.

The exhibit also anticipated the need for extensive highway networks capable of accommodating high-speed travel with optimized routing, an idea that later influenced the design of interstate highway systems and smart road technologies. Futurama's vision included dedicated lanes for autonomous vehicles, ensuring seamless transitions between urban and rural environments, reducing congestion, and enhancing overall traffic safety.

While purely conceptual at the time, this vision laid the foundation for key modern developments, including Vehicle-to-Infrastructure (V2I) communication technologies, smart road systems, and intelligent traffic management. Many of these principles continue to influence today's ongoing integration of autonomous vehicles into smart cities, shaping the future of intelligent transportation and mobility solutions.

1.3.2 1953: Wire-Guided Prototype

In 1953, RCA Labs pioneered one of the earliest practical implementations of autonomous vehicle technology by developing a wire-guided prototype. This vehicle operates by following electrical signals transmitted through wires embedded in the roadway, effectively steering itself along a predetermined path. This innovation represented a major leap forward in automation by demonstrating the feasibility of vehicles being externally controlled, eliminating the need for onboard decision-making. However, the reliance on external infrastructure also introduced significant limitations. The scalability of such systems was challenged by the high costs associated with embedding and maintaining specialized roadway infrastructure. Additionally, flexibility was restricted, as vehicles were unable to deviate from their predefined routes or adapt to unexpected obstacles. Despite these drawbacks, the RCA wire-guided system laid the foundation for future research into guided automation, influencing developments in magnetic and optical lane guidance systems, as well as early concepts of Vehicle-to-Infrastructure (V2I) communication. The core challenges identified in this early experiment—such as balancing infrastructure dependency with operational flexibility—continue to be key considerations in modern AV development.

1.3.3 1980s: Robotic Vans and DARPA Projects

The 1980s marked a turning point in autonomous vehicle development, with major strides being made in both the commercial and military sectors. Two significant areas of progress during this decade were the advancements made by Mercedes-Benz in robotic vehicle technology and the research efforts funded by the U.S. Defense Advanced Research Projects Agency (DARPA), both of which laid the foundation for modern AV systems.

Mercedes-Benz's Robotic Vans Mercedes-Benz was among the first automobile manufacturers to experiment with semi-autonomous driving systems. The company developed and tested a series of robotic vans equipped with an array of sensors and cameras that allowed the vehicles to recognize lane markings, detect obstacles, and interpret their surroundings in real time. These early prototypes made use of rudimentary computer vision techniques to analyze images captured from onboard cameras, an approach that would later evolve into the sophisticated deep learning-based perception models used in modern AVs. The experiments demonstrated the feasibility of computer-assisted lane detection and object recognition, proving that a vehicle could maintain its position on the road without continuous human control. This research not only showcased the potential of automated driving but also provided insights into the challenges of real-world road conditions, sensor limitations, and computational constraints.

DARPA Projects At the same time, DARPA was actively investing in autonomous vehicle research, funding projects that explored the use of LiDAR, radar, and AI-driven decision-making. These efforts were primarily aimed at developing self-driving military vehicles that could operate in challenging, unstructured environments without human intervention. DARPA-backed initiatives introduced the use of LiDAR for environmental mapping and pioneered some of the earliest AI-based path-planning algorithms.

One notable project was the Autonomous Land Vehicle (ALV), which combined machine vision with basic AI-driven navigation systems to traverse off-road terrain. While still in its infancy, the ALV laid the groundwork for the more advanced perception and control algorithms seen in today's autonomous military and civilian vehicles. The progress made through these projects helped shape future competitions such as the DARPA Grand Challenges in the 2000s, which would later play a crucial role in accelerating AV development.

By the end of the 1980s, the advancements made in both commercial and defense-related AV research demonstrated that automation was not just a theoretical concept but a viable technology that, with continued refinement, could one day revolutionize transportation.

1.3.4 1995: "No Hands Across America" [17]

In 1995, Carnegie Mellon University's NavLab project achieved a historic milestone in autonomous vehicle research with the "No Hands Across America" journey. This groundbreaking experiment involved a specially modified minivan, the NavLab 5, which successfully drove from Pittsburgh, Pennsylvania, to San Diego, California, covering nearly 3100 miles. What set this journey apart was that over 98% of the driving was handled autonomously, marking one of the earliest demonstrations of long-distance autonomous navigation on public roads.

NavLab 5 was equipped with an advanced sensor suite, including cameras and onboard computers that processed real-time visual and GPS data to make driving decisions. The vehicle used adaptive cruise control, lane-keeping assistance, and obstacle detection systems, which allowed it to navigate highways with minimal human intervention. While human operators were present in the vehicle for safety reasons, their role was largely supervisory, stepping in only when necessary.

The significance of this journey was immense. It provided invaluable data on how an autonomous vehicle could interact with real-world traffic conditions, adapt to changing environments, and make real-time adjustments based on sensor inputs. The project also demonstrated the potential of integrating GPS navigation with computer vision algorithms, a combination that remains fundamental in modern autonomous vehicle development. Moreover, "No Hands Across America" highlighted the practical challenges that AVs faced, such as handling urban environments, merging into high-speed traffic, and responding to unpredictable human drivers.

This milestone was not just a technical triumph but also a powerful demonstration of the possibilities of autonomous driving, inspiring further research and investment in the field. The success of NavLab 5 paved the way for future AV initiatives, including DARPA's Grand Challenges in the early 2000s, which accelerated advancements in artificial intelligence and robotic navigation. Today, the legacy of "No Hands Across America" can be seen in modern self-driving car projects that continue to build on the foundations laid by this pioneering achievement.

1.3.5 2004–2007: DARPA Grand Challenges

The DARPA Grand Challenges were a series of groundbreaking competitions organized by the U.S. Defense Advanced Research Projects Agency (DARPA) to accelerate the development of autonomous vehicles. These challenges played a pivotal role in advancing AV technology by fostering collaboration between academia, industry, and government institutions. By creating real-world obstacles and high-stakes competition, DARPA incentivized teams to push the boundaries of artificial intelligence, sensor integration, and path planning algorithms.

2004: The First Grand Challenge – A Harsh Reality Check The first DARPA Grand Challenge took place on March 13, 2004, in the Mojave Desert, where teams were tasked with developing autonomous vehicles capable of navigating a 142-mile off-road course without human intervention. The challenge highlighted the infancy of AV technology at the time, as none of the competing vehicles were able to complete the course. The best-performing vehicle, developed by Carnegie Mellon University, traveled only 7.4 miles before crashing. Major limitations included poor sensor reliability, insufficient computational capabilities, and ineffective decision-making algorithms. While the 2004 challenge did not yield a successful AV, it provided crucial insights into the weaknesses of existing technology and spurred significant improvements for the next competition.

2005: The Triumph of Stanley Building on the lessons from 2004, DARPA hosted a second challenge in 2005, featuring a 132-mile desert course with more refined technical requirements. This time, five vehicles successfully completed the race, showcasing substantial progress in AV design. The winning entry, "Stanley," developed by Stanford University, completed the course in 6 hours and 54 minutes using a combination of LiDAR, cameras, radar, and GPS-based navigation. Stanley's success was largely attributed to its innovative approach to perception and real-time path planning, using machine learning algorithms to adapt to unpredictable terrain conditions. This victory validated the feasibility of autonomous navigation in unstructured environments and marked a critical milestone in AV history.

2007: The Urban Challenge – Bringing AVs to City Streets The 2007 DARPA Urban Challenge took AV development to the next level by requiring vehicles to navigate a simulated urban environment. Unlike previous challenges that focused on off-road autonomy, the Urban Challenge introduced complexities such as obeying traffic laws, handling intersections, merging with other vehicles, and avoiding dynamic obstacles. The competition took place at a decommissioned military base designed to mimic real-world driving conditions, including multi-lane roads and intersections.

The winning vehicle, "Boss," developed by Carnegie Mellon University, successfully completed the 60-mile course in under six hours while demonstrating advanced decision-making capabilities, cooperative vehicle interactions, and real-time traffic negotiation. The Urban Challenge set the foundation for modern autonomous vehicle research by shifting the focus from isolated environments to real-world urban mobility.

Lasting Impact of the DARPA Grand Challenges The DARPA Grand Challenges were instrumental in catalyzing the AV industry, fostering innovations that later influenced commercial projects such as Google's self-driving car (later

Waymo), Tesla's Autopilot, and various automated transportation initiatives. Many of the researchers and engineers who participated in these challenges went on to work for leading AV companies, further propelling the field. Additionally, the challenges highlighted key technical hurdles—such as sensor fusion, decision-making under uncertainty, and real-time navigation—that continue to shape the development of AVs today. Ultimately, the success of the DARPA Grand Challenges demonstrated the viability of autonomous driving and laid the groundwork for the advancements that define the industry today.

- In 2004, none of the participating teams completed the desert course due to limitations in sensor range and decision-making algorithms.
- By 2005, Stanford's "Stanley" vehicle successfully navigated a 132-mile off-road course using advanced perception and planning algorithms.
- In 2007, DARPA introduced an Urban Challenge requiring vehicles to navigate city streets while obeying traffic laws—a significant leap toward real-world applications.

These competitions spurred innovation by fostering collaboration between academia and industry. Many technologies developed during these challenges form the basis of modern AV systems.

1.3.6 2010s: Commercialization Efforts

The 2010s marked a significant shift in autonomous vehicle development, transitioning from purely research-driven efforts to large-scale commercialization attempts. During this decade, major technology companies and automotive manufacturers began investing heavily in self-driving technology, aiming to bring autonomous systems to the consumer market. The key players in this era—Tesla, Waymo, and Uber—each took distinct approaches toward AV commercialization, showcasing both the potential and the challenges of autonomous driving.

Tesla: The Rise of Semi-autonomous Driving Tesla introduced its semi-autonomous Autopilot system in 2015, which leveraged a combination of cameras, radar, ultrasonic sensors, and AI-powered software to enable features like adaptive cruise control, lane-keeping assistance, and automatic lane changes. Tesla's strategy differed from fully autonomous projects in that it implemented a gradual, incremental approach, allowing vehicles to operate with increasing levels of automation while still requiring human oversight.

Tesla's system was marketed as a driver-assistance tool rather than a fully autonomous platform, yet it played a crucial role in popularizing autonomous driving among consumers. However, as Tesla pushed software updates enabling more advanced features, questions arose about over-reliance on automation and the limitations of its sensor suite, which lacked LiDAR technology. High-profile accidents involving Autopilot raised concerns about regulatory oversight, driver complacency, and the ethical implications of deploying semi-autonomous systems before achieving full reliability.

Waymo: Leading the Charge Toward Full Autonomy Originally a part of Google's self-driving car project, Waymo emerged as an independent company in 2016 under Alphabet Inc. Waymo took a more structured and safety-oriented approach compared to Tesla, focusing on developing fully autonomous vehicles (Level 4 and 5) that did not rely on human intervention. The company utilized a sensor-heavy approach, incorporating LiDAR, radar, high-definition mapping, and deep-learning AI to navigate complex urban environments.

Waymo's extensive testing in Phoenix, Arizona, led to the launch of the Waymo One ride-hailing service, the first fully autonomous taxi service available to the public. Unlike Tesla, which prioritized consumer-owned vehicles, Waymo pursued a shared mobility model, envisioning a future where private car ownership could be replaced by autonomous ride-hailing fleets. While Waymo demonstrated remarkable success in controlled urban environments, it also faced challenges related to scalability, high development costs, and difficulties in expanding to unpredictable road conditions in other cities.

Uber: The Challenges of Autonomous Ride-Sharing Uber entered the AV race with an ambitious goal: to develop self-driving taxis that could replace human drivers and revolutionize the ride-sharing industry. The company launched an autonomous vehicle testing program in Pittsburgh in 2016, outfitting Volvo SUVs with a suite of cameras, LiDAR, and AI-powered decision-making software. However, Uber's journey toward autonomy was marred by setbacks.

In 2018, an Uber self-driving test vehicle was involved in the first pedestrian fatality caused by an autonomous vehicle. This incident underscored the ethical and safety concerns of deploying self-driving technology before it had reached full maturity. Following the accident, Uber faced regulatory scrutiny, suspended its AV program for months, and later sold its autonomous vehicle division to Aurora Innovation in 2020. The incident highlighted the risks associated with AV deployment in uncontrolled environments and raised important questions about safety, accountability, and real-world readiness.

The Broader Impact of Commercialization Efforts

Despite these challenges, the commercialization efforts of the 2010s accelerated AV development and brought autonomous technology to the forefront of public awareness. Governments and regulatory bodies began drafting policies to govern self-driving technology, leading to discussions about safety standards, ethical dilemmas, and liability concerns. Additionally, the competition between companies fueled advancements in AI, sensor technology, and vehicle control systems.

While full autonomy remains an ongoing pursuit, the lessons learned from the 2010s continue to shape the future of AV technology, guiding current efforts to create safer, more efficient, and widely accepted self-driving systems.

1.4 Societal Impacts of Autonomous Vehicles

Autonomous vehicles (AVs) have the potential to fundamentally transform society, altering how people commute, how goods are transported, and how urban environments are designed. The integration of AVs into public and private transportation systems could lead to safer roads, reduced congestion, and improved accessibility. However, their widespread adoption also raises significant challenges, including ethical dilemmas, economic disruptions, and security concerns. Addressing these issues will require careful policymaking, advancements in technology, and a shift in public perception regarding automation in everyday life.

Beyond mobility, the ripple effects of AV technology extend to multiple sectors, including healthcare, logistics, and environmental sustainability. Autonomous vehicles can enhance emergency medical response times by optimizing ambulance routing, support supply chain efficiencies through autonomous freight transport, and contribute to greener cities by reducing unnecessary fuel consumption and emissions. Nevertheless, the road to full integration is paved with regulatory, legal, and infrastructural challenges that must be carefully navigated to maximize benefits while mitigating risks. This section explores the benefits and challenges AVs present and examines real-world case studies to understand their societal impact.

Autonomous vehicles (AVs) have the potential to fundamentally transform society, altering how people commute, goods are transported, and urban environments are designed. While the benefits of AVs are vast, including improved safety, efficiency, and accessibility, their deployment also raises significant challenges related to ethics, economy, and security. Understanding these societal impacts is crucial for policymakers, engineers, and businesses to navigate the future of autonomous mobility effectively.

Ethical and societal considerations are highlighted by Bonnefon et al. [7] and McCarroll [10], who investigate moral dilemmas and the necessity of experimental ethics in AV design. These themes are echoed in policy-oriented papers like that by Reimer and Nunes [15], which addresses the interplay between technology and regulatory frameworks.

Finally, validation and deployment are addressed by Yamazaki et al. [16], who stress the need for rigorous, scenario-based testing, ensuring safety before real-world implementation. The collective insight from these sixteen papers provides a multidimensional perspective essential for understanding the future trajectory of autonomous vehicle systems.

1.4.1 Benefits

AVs have the potential to significantly enhance safety by reducing accidents caused by human error—currently responsible for over 90% of traffic incidents. Additionally, they can improve efficiency by optimizing traffic flow and reducing fuel consumption. For individuals unable to drive due to age or disability, AVs offer newfound mobility and independence.

- Improved Safety: One of the most significant advantages of AVs is their potential to reduce traffic accidents caused by human error, which accounts for over 90% of crashes worldwide. By eliminating distractions, fatigue, and impaired driving, autonomous systems can significantly enhance road safety. AI-driven decision-making, coupled with advanced sensors such as LiDAR and radar, enables AVs to detect and respond to hazards more quickly than human drivers, reducing fatalities and injuries.
- Traffic Efficiency: AVs have the capability to optimize traffic flow by leveraging real-time data analytics and vehicle-to-vehicle (V2V) communication. By reducing unnecessary braking, improving lane discipline, and implementing coordinated vehicle movement, autonomous systems can minimize congestion and shorten travel times. Studies indicate that a higher percentage of AVs on the road could lead to a substantial reduction in traffic bottlenecks and carbon emissions from idling vehicles.
- Accessibility and Mobility Expansion: Autonomous technology has the potential to provide enhanced mobility for individuals who are elderly, disabled, or otherwise unable to drive. Self-driving cars can offer independence to these populations, improving their ability to participate in economic and social activities. Additionally, AVs could transform public transportation by providing on-demand, efficient transit services in areas where traditional options are limited.
- Environmental Benefits: AVs can contribute to sustainability efforts by optimizing acceleration, braking, and route selection, leading to improved fuel efficiency and reduced emissions. Additionally, the shift to electric autonomous fleets could further decrease the reliance on fossil fuels, promoting a cleaner and more energy-efficient transportation system.

1.4.2 Challenges

However, AVs also present ethical dilemmas in decision-making during life-threatening situations (e.g., prioritizing passenger safety over pedestrians). Economic disruptions are another concern; industries reliant on human drivers may face job displacement as automation becomes widespread. Finally, data privacy issues arise from the vast amounts of information collected by AV systems.

- Ethical Dilemmas and Decision-Making Challenges: AVs must be programmed to handle critical ethical decisions in unavoidable accident scenarios. The classic "trolley problem"—where a vehicle must choose between harming different individuals—illustrates the complexity of ethical programming in AVs. These moral dilemmas raise questions about liability, accountability, and public acceptance of autonomous decision-making.

- Economic Disruptions and Workforce Displacement: The widespread adoption of AVs threatens traditional transportation jobs, particularly for truck drivers, taxi operators, and delivery personnel. While AVs may create new roles in vehicle maintenance, data analysis, and AI development, the transition could lead to significant workforce disruptions. Governments and industries must develop strategies to retrain affected workers and mitigate economic displacement.
- Privacy and Security Concerns: AVs generate and process vast amounts of data, including location tracking, passenger behavior, and real-time environmental conditions. This data collection raises concerns about privacy and potential misuse. Additionally, the cybersecurity of AVs is a critical issue—hacking or system malfunctions could pose severe safety risks. Robust encryption, regulatory oversight, and cybersecurity protocols must be developed to ensure the security of AV networks.
- Infrastructure and Legal Adaptation: AV adoption requires significant upgrades to existing infrastructure, including smart traffic signals, dedicated AV lanes, and vehicle-to-infrastructure (V2I) communication systems. Additionally, legal and regulatory frameworks must evolve to address liability in accidents involving AVs, setting clear guidelines on manufacturer responsibility and user accountability.

Case Studies and Real-World Implementations

Several pilot programs have tested AVs in real-world conditions, revealing both promising benefits and operational challenges:

Waymo One in Phoenix, Arizona: This fully autonomous ride-hailing service has demonstrated that AVs can operate safely in controlled environments, but scaling the model to diverse geographies remains a challenge.
Autonomous Shuttle Trials in Europe: Cities like Paris and Helsinki have experimented with self-driving shuttles for public transportation. While these systems have improved mobility options, they have also highlighted challenges in integrating AVs with traditional transit networks.
Tesla's Full Self-Driving (FSD) Beta Program: Tesla's FSD system has showcased the potential for consumer-owned AVs, but regulatory scrutiny and high-profile incidents continue to shape public perception and industry oversight.

These case studies emphasize the need for continued research, policy refinement, and public engagement to maximize the benefits of AVs while mitigating their societal risks.

1.5 Technical Challenges in Perception, Planning, and Control

Developing reliable autonomous vehicles (AVs) requires addressing significant technical hurdles across multiple domains, including perception, planning, and control. These challenges stem from the complexity of real-world driving environments, where AVs must continuously interpret dynamic surroundings, make rapid decisions, and execute precise maneuvers. Unlike controlled laboratory settings, real-world roads present unpredictable variables such as erratic human drivers, diverse weather conditions, road obstructions, and unexpected hazards. Overcoming these technical challenges is crucial for ensuring the safety, efficiency, and widespread adoption of AVs.

1.5.1 Perception

Perception is the foundation of autonomous driving, allowing AVs to interpret and understand their surroundings. An AV's perception system relies on multiple sensors, including cameras, LiDAR, radar, and ultrasonic sensors, to detect objects, road conditions, and traffic signals. However, perception presents several challenges:

- Sensor Limitations: While LiDAR provides high-resolution 3D mapping, it struggles in adverse weather conditions such as heavy rain, fog, and snow, where reflections can distort sensor readings. Similarly, cameras are essential for visual recognition but are highly susceptible to lighting variations, including glare and low-light conditions. Radar, although more resilient in inclement weather, has lower spatial resolution and may misinterpret small objects.
- Occlusions: Objects hidden by other vehicles, pedestrians, or environmental structures can pose a significant challenge. For example, an AV may need to anticipate a pedestrian crossing behind a parked truck or detect a cyclist emerging from an alleyway. Advanced sensor fusion techniques and predictive modeling are being developed to compensate for occlusions and enhance situational awareness.
- Real-Time Processing: AVs must process vast amounts of sensor data in real time to make split-second decisions. Achieving this requires high-performance computing capabilities, edge computing solutions, and optimized machine learning models capable of handling complex environments without latency-induced delays. Efficient data compression, parallel processing architectures, and hardware acceleration (e.g., GPUs and TPUs) are critical for meeting these demands.
- Semantic Understanding: Beyond basic object detection, AVs must interpret the intentions and behaviors of surrounding entities. For instance, recognizing a pedestrian looking at their phone near a crosswalk might indicate they are likely

to cross inattentively. Similarly, identifying a driver aggressively changing lanes could signal an increased risk of erratic movement.

1.5.2 Planning

Once an AV has perceived its environment, it must determine the optimal path to reach its destination safely and efficiently. Planning involves both high-level route selection and low-level motion planning to navigate complex environments. Key challenges in planning include:

- Unpredictable Obstacles: Roads are filled with unpredictable elements such as erratic drivers, sudden braking, jaywalking pedestrians, and debris. AVs must employ probabilistic models and reinforcement learning techniques to adapt dynamically to these uncertainties while maintaining safety.
- Dynamic Traffic Conditions: Unlike pre-planned routes used in GPS navigation, AVs must continuously update their routes based on real-time traffic data, road closures, and temporary construction zones. This requires integrating live data streams, including vehicle-to-vehicle (V2V) and vehicle-to-infrastructure (V2I) communication, to make informed routing decisions.
- Real-Time Decision-Making: Unlike traditional rule-based driving, AVs must weigh multiple factors when deciding their next action. For example, when merging onto a highway, an AV must assess the speed and behavior of surrounding vehicles, predict gaps in traffic, and execute the merge smoothly without disrupting other drivers. Deep reinforcement learning algorithms, inverse reinforcement learning, and Monte Carlo tree search techniques are being developed to refine AV decision-making capabilities.
- Ethical Considerations: In unavoidable accident scenarios, AVs must determine the best course of action with minimal harm. The ethical programming of AVs remains a controversial topic, as different stakeholders (manufacturers, regulators, and the public) have varying opinions on how AVs should prioritize safety in complex scenarios.

1.5.3 Control

Control systems ensure that an AV executes its planned trajectory safely and efficiently. The control layer translates planned actions into precise vehicle movements, adjusting acceleration, braking, and steering in real time. Major challenges in AV control include:

- Vehicle Stability: AVs must maintain stability under diverse conditions, including high-speed highway driving, sharp turns, and emergency maneuvers. Advanced control algorithms, such as Model Predictive Control (MPC) and Proportional-

Integral-Derivative (PID) controllers, are employed to ensure smooth and stable operation.
- Actuator Response Time: The ability of an AV's mechanical systems (e.g., brakes, steering, and throttle) to respond accurately to electronic commands is crucial. Any delay in response could compromise safety, especially in critical situations requiring sudden braking or evasive maneuvers.
- Fail-Safe Mechanisms: System failures, whether due to sensor malfunctions, software bugs, or unexpected environmental conditions, must be managed through robust fail-safe mechanisms. Redundant sensor architectures, failover control algorithms, and emergency stop protocols are essential to maintaining operational reliability.
- Adapting to Road Surface Conditions: AVs must dynamically adjust their driving style based on road conditions such as wet, icy, or uneven surfaces. Real-time traction control and adaptive suspension systems help mitigate risks associated with variable terrains.

1.6 Ethical and Societal Reflection Exercises

Autonomous Vehicles (AVs) bring not only technological but also ethical challenges. Each of the following scenarios presents a moral dilemma relevant to AV design, regulation, or deployment. Read each case carefully and write a response (150–250 words) that addresses the guiding questions.

As we proceed through the rest of this book, we will delve into the technical foundations that power AV systems from sensor integration and perception to path planning and control. However, understanding the technology without grappling with its ethical and societal implications would be incomplete. These questions are not merely academic; they will shape public trust, legal frameworks, and the very fabric of how AVs coexist with human beings. By reflecting on these dilemmas early, students and practitioners can develop a mindset that balances innovation with responsibility, and learn to design systems that are not only intelligent but also just, inclusive, and accountable.

1.6.1 Scenario 1: Decision at the Crosswalk

An AV is driving down a two-lane urban road. Suddenly, a pedestrian jaywalks into the vehicle's path. The vehicle has only two options:

- Continue forward, likely injuring the pedestrian.
- Swerve into a concrete barrier, risking serious injury to the passenger.

Questions
- What action should the vehicle take in this scenario? Why?
- What ethical framework are you using to justify your choice?
- How should vehicle manufacturers or regulators handle such dilemmas?

1.6.2 Scenario 2: The Unavoidable Collision

An AV is driving on a narrow mountain road. A child runs into the path. On the other side is a steep cliff. The car must either hit the child or drive off the road.

Questions
- What should the AV be programmed to do?
- How would you justify that ethically?
- Should passengers be able to override such decisions?

1.6.3 Scenario 3: Economic Disruption

AVs may lead to mass unemployment among professional drivers.

Questions
- Should governments or companies slow AV deployment to protect jobs?
- What responsibilities do AV companies have toward displaced workers?
- Can you think of a historical example of similar disruption?

1.6.4 Scenario 4: Data Privacy in AVs

AVs continuously collect data from users and the environment.

Questions
- Who should own and control this data?
- Should users be able to opt out of data collection?
- How should privacy differ between personal AVs and public fleets?

References

1. Bimbraw, K. (2015). Autonomous cars: Past, present and future a review of the developments in the last century, the present scenario and the expected future of autonomous vehicle technology. In *2015 12th International Conference on Informatics in Control, Automation and Robotics (ICINCO)* (Vol. 1, pp. 191–198). IEEE.

2. Chen, L., Li, Y., Huang, C., Li, B., Xing, Y., Tian, D., Li, L., Hu, Z., Na, X., & Li, Z. (2022). Milestones in autonomous driving and intelligent vehicles: Survey of surveys. *IEEE Transactions on Intelligent Vehicles*, *8*(2), 1046–1056.
3. Chen, M., Jochem, T., & Pomerleau, D. (1995). Aurora: A vision-based roadway departure warning system. In *Proceedings 1995 IEEE/RSJ International Conference on Intelligent Robots and Systems. Human Robot Interaction and Cooperative Robots* (Vol. 1, pp. 243–248). IEEE.
4. Dong, Z., Shi, W., Tong, G., & Yang, K. (2020). Collaborative autonomous driving: Vision and challenges. In *2020 International Conference on Connected and Autonomous Driving (MetroCAD)* (pp. 17–26). IEEE.
5. Huang, Y., & Chen, Y. (2020). *Autonomous driving with deep learning: A survey of state-of-art technologies*. Preprint. arXiv:2006.06091
6. Khan, M. A., Sayed, H. E., Malik, S., Zia, T., Khan, J., Alkaabi, N., & Ignatious, H. (2022). Level-5 autonomous driving–are we there yet? A review of research literature. *ACM Computing Surveys (CSUR)*, *55*(2), 1–38.
7. Klaver, F. (2020). The economic and social impacts of fully autonomous vehicles. *Retrieved February*, *18*, 2022.
8. Liu, L., Lu, S., Zhong, R., Wu, B., Yao, Y., Zhang, Q., & Shi, W. (2021). Computing systems for autonomous driving: State of the art and challenges. *IEEE Internet of Things Journal*, *8*(8), 6469–6486. ID: 1.
9. Liu, S., et al. (2019). Edge computing for autonomous driving: Opportunities and challenges. *Proceedings of the IEEE*, *107*(8), 1697–1716.
10. McCarroll, C., & Cugurullo, F. (2022). Social implications of autonomous vehicles: A focus on time. *AI and Society*, *37*(2), 791–800.
11. Omeiza, D., Webb, H., Jirotka, M., & Kunze, L. (2021). Explanations in autonomous driving: A survey. *IEEE Transactions on Intelligent Transportation Systems*, *23*(8), 10142–10162.
12. Rosenzweig, J., & Bartl, M. (2015). A review and analysis of literature on autonomous driving. *E-Journal Making-of Innovation*, 1–57.
13. SAE. (2021). Taxonomy and definitions for terms related to driving automation systems for on-road motor vehicles. https://new.sae.org/standards/j3016_202104-taxonomy-definitions-terms-related-driving-automation-systems-road-motor-vehicles
14. Trypuz, R., Kulicki, P., & Sopek, M. (2024). Ontology of autonomous driving based on the SAE J3016 standard. *Semantic Web*, *15*(5), 1837–1862.
15. Wang, Y., Liu, L., Zhang, X., & Shi, W. (2019). Hydraone: An indoor experimental research and education platform for CAVs. In *2nd USENIX Workshop on Hot Topics in Edge Computing (HotEdge 19)*.
16. Woo, S., Youtie, J., Ott, I., & Scheu, F. (2021). Understanding the long-term emergence of autonomous vehicles technologies. *Technological Forecasting and Social Change*, *170*, 120852.
17. worldsfairphotos.com. (2022). *The 1964–1965 New York world's fair*.
18. Yurtsever, E., Lambert, J., Carballo, A., & Takeda, K. (2020). A survey of autonomous driving: Common practices and emerging technologies. *IEEE Access*, *8*, 58443–58469.

Chapter 2
Simulation Playground

2.1 Introduction

Simulation environments are the backbone of autonomous vehicle (AV) development, providing a safe, cost-effective platform to test and refine algorithms. This chapter focuses on setting up simulation environments tailored for different operating systems and deployment scenarios. Additionally, it provides detailed explanations of key scripts, such as manual control and traffic generation, to help students and lecturers gain hands-on experience with simulation tools like CARLA and BlueICE.

CARLA (Car Learning to Act) is an open-source simulator specifically designed for autonomous driving research. Developed to provide high-fidelity simulation environments, CARLA supports configurable weather, traffic, and sensor models—making it ideal for testing perception, planning, and control algorithms under varied conditions. As the simulation backbone of the BlueICE framework used in this book, CARLA enables students and researchers to safely explore complex driving scenarios in a repeatable and scalable virtual environment. Its compatibility with Python and ROS enhances accessibility and integration with real-world autonomous driving stacks.

2.2 BlueICE

To support hands-on learning throughout this book, we will use the BlueICE framework as the primary simulation environment. As shown in Fig. 2.1, BlueICE enables a modular and distributed approach to autonomous vehicle (AV) simulation. Students will use this framework to explore how perception, planning, and control components behave under realistic, complex conditions. Unlike traditional simulators that operate on a single machine, BlueICE distributes tasks—such as vehicle

W. Shi, Y. He, *Introduction to Autonomous Driving*,
https://doi.org/10.1007/978-3-031-99485-2_2

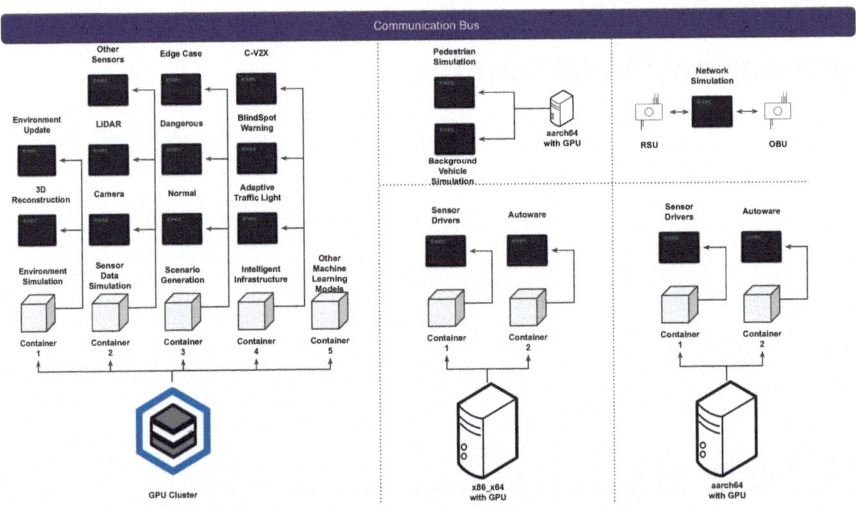

Fig. 2.1 BlueICE framework co-simulation overview. The digital twin uses different computing equipment to host and run different simulation applications. The communication bus combines the different simulators together to orchestrate and synchronize data [1]

dynamics, sensor processing, and traffic coordination—across multiple computing nodes, allowing for high-fidelity experimentation at scale.

At its core, BlueICE is designed for extensibility and interactivity. Each simulator runs inside its own container, making it easy to isolate environments, manage dependencies, and incorporate new tools. Communication between simulation components is handled through a hybrid messaging system: ROS2 is used for high-speed, low-latency messaging in local deployments, while MQTT supports broader cloud-based synchronization. As students begin working with BlueICE, they will gain direct experience integrating these communication protocols and observing how real-time data flows across a co-simulation architecture.

What makes BlueICE especially valuable for education is its deterministic synchronization mechanism, which ensures that all simulators operate in lockstep. This is critical when coordinating AV behavior across distributed components. Rather than allowing simulators to drift apart in time, BlueICE uses a global Simulation Manager that controls when each node can advance to the next frame. This mechanism provides students with a predictable, reproducible simulation environment—essential for debugging, performance testing, and system validation.

BlueICE is also compatible with a wide range of simulation backend, including CARLA-based co-simulators. Instructors can flexibly configure which modules are active during a session, allowing each classroom setup to be tailored to the lesson's goals. Furthermore, BlueICE supports several digital twin environments, which simulate real-world infrastructure and campus layouts. A list of supported maps and locations can be found on the BlueICE GitHub repository.[1] Students will use these environments to simulate real-time interactions between AVs and connected infrastructure—an essential experience for understanding how autonomous systems behave in smart cities and edge-assisted deployments.

2.3 Setting Up BlueICE

BlueICE's central simulation backbone is CARLA. This section describes how to install and configure CARLA as a standalone simulator, and then how to prepare it for use within the BlueICE framework. BlueICE layers orchestration and integration with Autoware on top of it. Proper setup ensures both components operate together reliably and efficiently. Below is a platform-specific guide:

2.3.1 *Windows Setup*

1. Download the CARLA simulator from its official GitHub release page.
2. Install Python 3.7 or higher and ensure it is added to the system PATH.
3. Extract the CARLA archive to a directory (e.g., `C:\CARLA`).
4. Launch CARLA by running `CarlaUE4.exe`.
5. Open a command prompt and navigate to the `PythonAPI\util` directory. Run `config.py` to verify Python dependencies.
6. Install required Python packages using `pip install -r requirements.txt`.

2.3.2 *Linux Setup*

1. Download the Linux CARLA build from the GitHub releases.
2. Install Python 3.x and pip: `sudo apt install python3 python3-pip`.
3. Extract files to a directory (e.g., `~/CARLA`).
4. Run `./CarlaUE4.sh` to start the simulator.
5. Run `python3 PythonAPI/util/config.py` to check setup and install any missing dependencies.

[1] BlueICE Digital Twin Maps: https://github.com/Croquembouche/BlueICE.

2.3.3 macOS Setup

1. Support is limited; refer to CARLA's documentation for the latest updates.
2. Install Python using Homebrew: `brew install python`.
3. Clone and build the CARLA repository from source.
4. Navigate to the CARLA directory and run `./CarlaUE4.sh`.
5. Test using an example like `manual_control.py` in the PythonAPI examples folder.

For greater flexibility in customizing maps and traffic agents, instructors are strongly encouraged to use a source-built version of CARLA to set up the simulation environment for the course.

2.3.4 Example CARLA Script Usage

CARLA provides Python scripts for manual control, traffic generation, and weather simulation. These are helpful in familiarizing yourself with the API before integrating Autoware.

2.3.5 manual_control.py: Allows Keyboard or Joystick Control of an Ego Vehicle

```
# Connect to the CARLA server at localhost on port 2000
client = carla.Client('localhost', 2000)
client.set_timeout(10.0) # Set a timeout in seconds for server responses

# Retrieve the simulation world object
world = client.get_world()

# Get access to all blueprints (vehicle and sensor definitions)
blueprint_library = world.get_blueprint_library()

# Select a specific vehicle blueprint, in this case a Tesla Model 3
vehicle_bp = blueprint_library.filter('model3')[0]

# Choose a spawn point from the map
spawn_point = world.get_map().get_spawn_points()[0]

# Spawn the vehicle actor at the chosen location
vehicle = world.spawn_actor(vehicle_bp, spawn_point)

# Enable autopilot mode so the vehicle can drive autonomously
vehicle.set_autopilot(True)
```

2.3.6 generate_traffic.py: Spawns Multiple Vehicles and Pedestrians for Testing Perception Systems

```
# Get all possible spawn points from the map
spawn_points = world.get_map().get_spawn_points()

# Filter all available vehicle blueprints
vehicle_bps = blueprint_library.filter('vehicle.*')

# Spawn 10 non-player vehicles (NPCs)
for i in range(10):
    transform = spawn_points[i % len(spawn_points)] # Cycle through spawn
        points
    bp = random.choice(vehicle_bps) # Randomly select a vehicle blueprint
    npc = world.try_spawn_actor(bp, transform) # Try spawning the vehicle
    if npc:
        npc.set_autopilot(True) # Enable autopilot for NPC vehicles
```

2.3.7 weather_control.py: Dynamically Changes Simulation Weather Conditions

```
# Define a weather setting with clouds, rain, and fog
weather = carla.WeatherParameters(
    cloudiness=80.0,
    precipitation=60.0,
    fog_density=20.0
)

# Apply the weather setting to the simulation world
world.set_weather(weather)
```

These scripts provide a foundational toolkit to explore how different components in the simulation environment interact.

2.4 Integrating Autoware with BlueICE

One of BlueICE's strongest capabilities is its integration with Autoware. This connection uses containerized ROS 2 nodes and bridging utilities to enable end-to-end AV simulation.

The **AutowareUniverse-Carla** repository bridges CARLA and Autoware, enabling sensor data simulation and communication with Autoware's perception and control pipelines.[2]

2.4.1 *host_vehicle.py: Full ROS 2 Node Script for Managing Ego Vehicle in CARLA*

```
# This script sets up and manages an ego vehicle in the CARLA simulator
    using ROS 2.
# It enables sensor integration, control via Autoware or manual ROS 2
    commands,
# and publishes vehicle state information to ROS 2 topics.

import rclpy
from rclpy.node import Node
from sensor_msgs.msg import Imu, Image, PointCloud2
from geometry_msgs.msg import PoseWithCovarianceStamped
from autoware_auto_vehicle_msgs.msg import VelocityReport, SteeringReport,
    ControlModeReport
from autoware_auto_control_msgs.msg import AckermannControlCommand
from std_msgs.msg import String
import carla

class HostVehicle(Node):
    def __init__(self):

        #Connect to CARLA and Prepare CARLA World Variables
        self.connectCARLA()
        self.setUpPublishers()
        self.setUpSubscribers()

    def setUpPublishers(self):
        # Create publishers for vehicle state and sensor outputs
        # These topics are useful for monitoring and Autoware integration
        self.gnss_pub = self.create_publisher(PoseWithCovarianceStamped, "/
            sensing/gnss/pose_with_covariance", 10)
        self.imu_pub = self.create_publisher(Imu, "/sensing/imu/imu_data",
            10)
        self.vehicle_report_pub = self.create_publisher(VelocityReport, "/
            vehicle/status/velocity_status", 10)
        self.steering_report_pub = self.create_publisher(SteeringReport, "/
            vehicle/status/steering_status", 10)
        self.vehicle_status_control_mode_pub = self.create_publisher(
            ControlModeReport, "/vehicle/status/control_mode", 10)
```

[2] Full code and updates available at: https://github.com/Croquembouche/AutowareUniverse-Carla/tree/main.

```
    def setUpSubscribers(self):
        # Subscribe to Autoware control commands for autonomous navigation
        self.create_subscription(AckermannControlCommand,
                                 "/control/trajectory_follower/control_cmd",
                                 self.AutowareControllCallbacks, 10)

        # Subscribe to manual control commands published as plain strings
        self.create_subscription(String, 'hv_controls', self.
            vehicleROSControls, 10)

    def AutowareControllCallbacks(self, controlmsg):
        # Handle control commands from Autoware
        desired_steering = controlmsg.lateral.steering_tire_angle # Read
            target steering angle
        desired_speed = controlmsg.longitudinal.speed # Read target forward
            speed

        # Update vehicle's steering angle (negated due to coordinate
            convention)
        self.ego_control.steer = -desired_steering

        # Apply braking or throttle based on desired speed
        if desired_speed == 0:
            self.ego_control.brake = 1 # Full stop
        else:
            self.ego_control.throttle = 0.5 # Apply half throttle
            self.ego_control.brake = 0

        # Send the control command to the simulated vehicle
        self.vehicle.apply_control(self.ego_control)

    def vehicleROSControls(self, msg):
        # Enable or disable autopilot using a string command
        if msg.data == "auto␣on":
            self.vehicle.set_autopilot(True)
        elif msg.data == "auto␣off":
            self.vehicle.set_autopilot(False)

# Main function: initializes ROS 2 and starts the node

def main(args=None):
    rclpy.init(args=args)
    host_veh = HostVehicle() # Create node instance
    try:
        rclpy.spin(host_veh) # Keep node running
    except KeyboardInterrupt:
        host_veh.destroy_node() # Clean shutdown on Ctrl+C
        rclpy.shutdown()

if __name__ == '__main__':
    main()
```

This integration supports full autonomy workflows, from perception to planning and control, within a digital twin environment. It allows repeatable experiments and system debugging before any real-world deployment.

2.5 Autoware.Universe

Autoware.Universe is a comprehensive, modular software platform developed as part of the Autoware project to enable full-stack autonomous driving capabilities using ROS 2. It consolidates a wide array of autonomous vehicle components—including perception, localization, planning, control, and vehicle interface—into a unified framework designed for extensibility and real-world deployment. Autoware.Universe emphasizes flexibility by supporting diverse vehicle and sensor configurations through parameterized modules and standardized interfaces. Each functional domain is organized as a standalone component that can be selectively enabled or customized, making the system highly adaptable for simulation as well as real-vehicle applications. The platform integrates modern software practices such as containerization, multi-threaded processing, and standardized APIs, which facilitate scalable development and deployment. Built atop the foundational ROS 2 middleware, Autoware.Universe also ensures robust inter-process communication and compatibility with emerging standards in autonomous systems engineering.

Sections of the Autoware node diagram[3] are selected for additional review (Fig. 2.2).

2.5.1 Sensing

The sensing stack begins with raw point cloud data from the LiDAR sensor. This data first enters the `crop_box_filter_self`, which removes points that correspond to the ego vehicle's body to prevent false object detections from self-reflections. Next, the `crop_box_filter_mirror` excludes points likely originating from reflective surfaces such as mirrors or glass, which can distort object perception. After spatial filtering, the `distortion_corrector` module compensates for motion-induced warping of LiDAR scans by using synchronized IMU and velocity data. This correction is essential when the vehicle is in motion during LiDAR spinning. The `ring_outlier_filter` is then applied to remove sparse or misaligned points that deviate from expected ring patterns in multi-beam LiDAR data. Finally, the cleaned point cloud proceeds through `voxel_grid_filter` and

[3] Latest Autoware node diagram: https://autowarefoundation.github.io/autoware-documentation/main/design/autoware-architecture/node-diagram/.

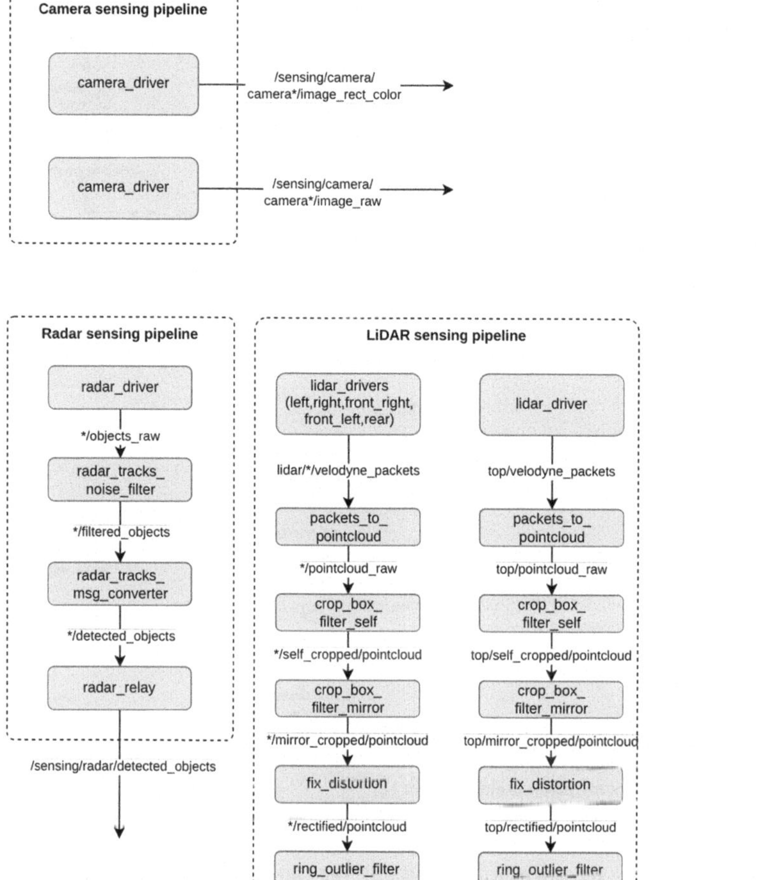

Fig. 2.2 Autoware Sensing node diagram[2]

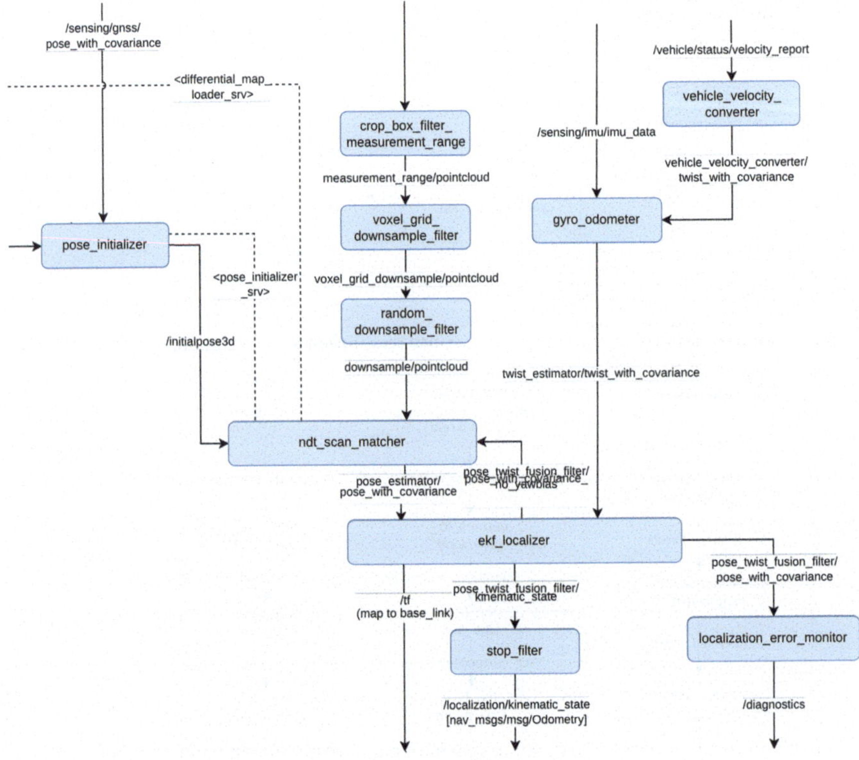

Fig. 2.3 Autoware Localization node diagram [2]

`distance_filter` to downsample and clip the range of the data, optimizing it for real-time use in the perception stack (Fig. 2.3).

2.5.2 *Localization*

Autoware's localization system performs sensor fusion to estimate the ego vehicle'ffs position and orientation in a global reference frame. It integrates data from GNSS, IMU, and LiDAR sources. `ekf_localizer` acts as a central filter, typically an Extended Kalman Filter, which ingests velocity data, acceleration, and GNSS fixes to compute an initial pose estimate. LiDAR-based localization via `ndt_scan_matcher` aligns incoming point clouds with a high-definition map to refine this estimate. The output is then corrected for biases and frame alignment using `pose_initializer` and `transform_listener`, resulting in a highly accurate, globally consistent pose. This output is published as `/localization/pose_twist_fusion`, which feeds directly into both global and local planning systems (Fig. 2.4).

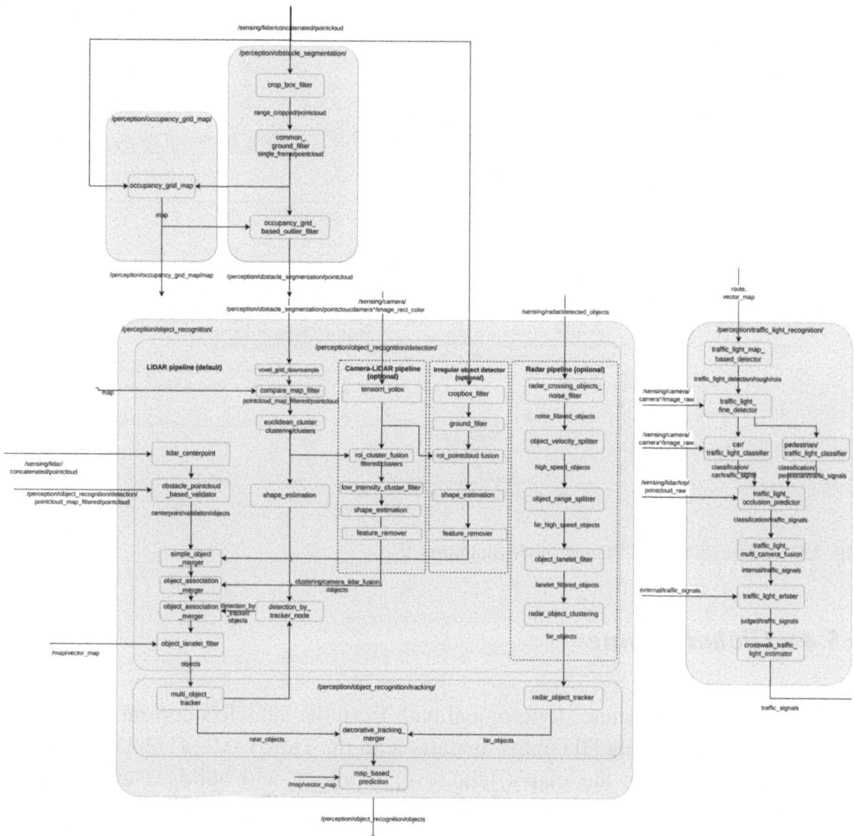

Fig. 2.4 Autoware Perception node diagram [2]

2.5.3 *Perception*

The perception stack ingests the filtered point cloud and transforms it into a semantic understanding of the environment. The `ray_ground_classifier` separates ground points from non-ground points, which simplifies the task of object detection. `euclidean_cluster` is then used to group non-ground points into candidate objects using spatial proximity. These clusters are passed to modules like `object_map_fusion`, which combines sensor observations with prior map knowledge, and `object_merger`, which consolidates overlapping detections from multiple sensor modalities. Parallel to the LiDAR stream, the camera input is processed by object detection models like `tensorrt_yolox` for 2D bounding boxes, and `multi_object_tracker` maintains temporal consistency across frames. These results are fused via `fusion_node` to produce a unified object list, containing tracked, classified, and spatially located objects. This perception output is critical for both planning and prediction stages (Fig. 2.5).

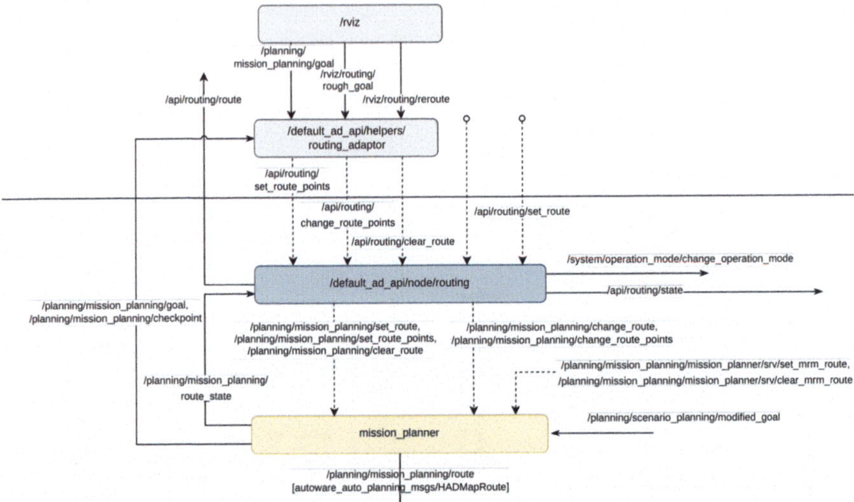

Fig. 2.5 Autoware Global Planner node diagram [2]

2.5.4 *Global Planner*

The global planner creates a topological path from the vehicle's current location to a user-defined goal using HD maps. It starts with the `lanelet2_global_planner` node, which interprets the map's lane-level topology and builds a graph representation of drivable paths. Once a route is selected based on routing constraints and vehicle capabilities, the planner generates a series of waypoints forming the global path. These waypoints are annotated with speed limits, intersection flags, and turn instructions. The result is a high-level navigational plan that avoids illegal maneuvers and remains entirely on mapped roads. This global path serves as a reference corridor for downstream local planning and behavior modules (Fig. 2.6).

2.5.5 *Local Planner*

Local planning refines the global path into a feasible, safe, and smooth trajectory. The process begins with scenario selection through `scene_module_manager`, which dynamically activates modules like `intersection`, `lane_change`, or `pull_over` based on the current driving context. The `behavior_path_planner` then generates candidate trajectories that account for nearby objects, road geometry, and dynamic constraints. These are further processed by `motion_velocity_smoother`, which regulates acceleration and deceleration profiles for passenger comfort and safety. Local planning outputs a time-parametrized trajectory that

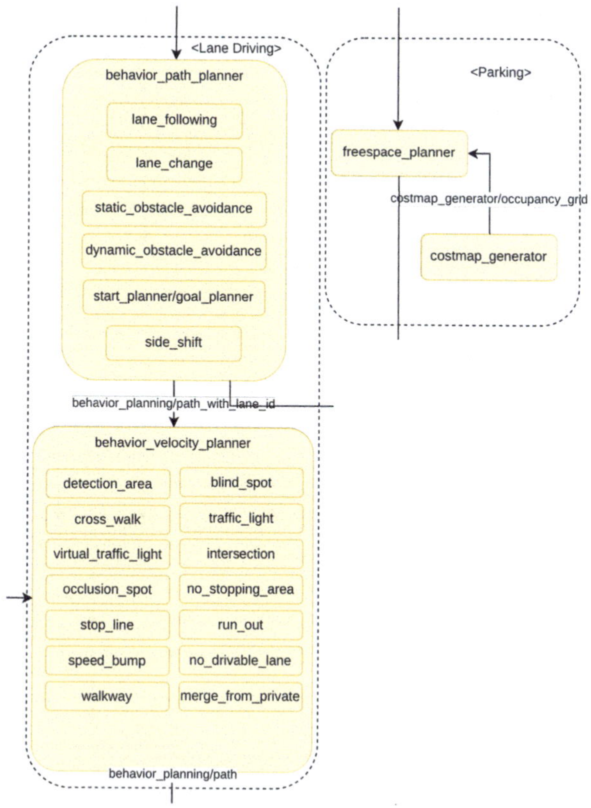

Fig. 2.6 Autoware Local Planner node diagram [2]

maintains alignment with the global route while adapting to real-time conditions (Figs. 2.7 and 2.8).

2.5.6 *Motion Smoother and Control Gate*

Once the trajectory is generated, it is passed to `control_command_publisher`, which translates it into low-level commands: target speed, steering angle, and brake force. This passes through `control_validator`, which checks the trajectory against control limits and system safety constraints. The validated command stream then reaches the `control_gate`, a supervisory node that manages transitions between autonomous and manual control modes. If autonomy is disengaged or a failure is detected, the gate can suppress commands or hand over control to a human driver. Finally, `vehicle_interface` converts these commands into platform-

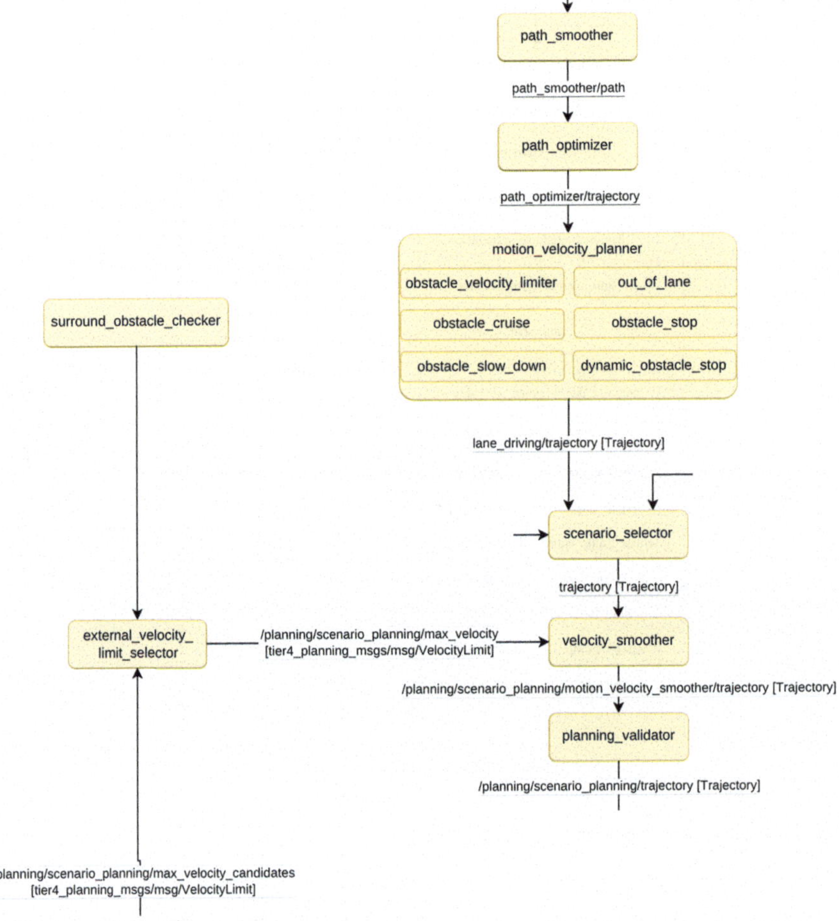

Fig. 2.7 Autoware Motion Smoother node diagram [2]

specific signals (e.g., CAN messages) and sends them to the actual vehicle through the `pacmod_interface` or `vehicle_cmd_gate`.

With the correct setup, Autoware can be an extremely powerful education tool for new and experienced programmers. While we do not dive into the details of how Autoware functions here, this chapter provides a link to the official documentation[4] and a docker image and Autoware.Universe code that is ready to use.[5]

[4] Official Autoware Documentation: https://autowarefoundation.github.io/autoware-documentation/main/.

[5] Docker Container and Source Code: https://github.com/Croquembouche/UDAutoware.

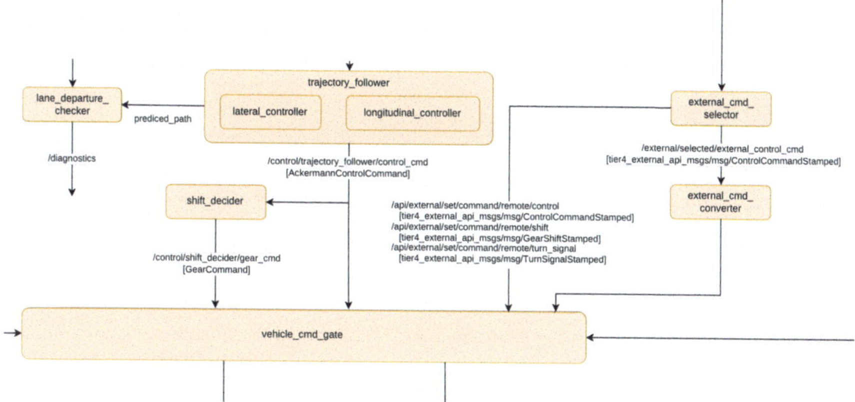

Fig. 2.8 Autoware Control Gate node diagram [2]

2.6 Hands-on Exercises

2.6.1 Run Autoware.Universe Using BlueICE and the AutowareCarla Bridge

Task Deploy a modular autonomous driving stack using Autoware.Universe within the BlueICE simulation platform. Launch the ego vehicle in CARLA, bridge sensor data to Autoware via ROS 2, and observe full-stack operation including perception, localization, and control.

Object Use Docker-based BlueICE and Autoware-Carla setup. Start CARLA simulation, run host vehicle.py to initialize the ego vehicle, and launch Autoware nodes. Monitor system output using RViz and ROS 2 topic introspection.

Deliverable Submit a short report with:

- Diagram of your setup and data flow.
- RViz screenshot showing perception and localization.
- Short video or screenshots of the ego vehicle in autonomous mode.
- Brief summary of key ROS topics used.

Optional Include one integration issue encountered and how it was resolved.

References

1. He, Y., Chen, H., Shi, W. (2024). *An advanced framework for ultra-realistic simulation and digital twinning for autonomous vehicles*. Preprint. arXiv:2405.01328.
2. Kato, S., et al. (2015). An open approach to autonomous vehicles. *IEEE Micro*, *35*(6), 60–68.

Chapter 3
Sensor Technologies

3.1 Introduction

Sensors are foundational components in autonomous systems, serving as the vehicle's means of perceiving and interpreting the external environment, analogous to human sensory systems. All autonomous driving algorithms rely on the integrity and quality of sensor-derived data. Figure 3.1 illustrates the principal sensors commonly used in autonomous vehicles. It is important to note that these are specialized for autonomous driving and exclude standard automotive sensors such as those monitoring voltage, airbag deployment, or tire pressure.

For clarity, these sensors can be categorized into two types: **exteroceptive sensors** and **proprioceptive sensors** [3]. Exteroceptive sensors acquire data from the vehicle's surroundings, while proprioceptive sensors provide information about the vehicle's internal state and dynamics. Each category is detailed in the following sections.

3.2 Exteroceptive Sensors

Exteroceptive sensors are essential for perceiving the external environment and enabling safe navigation. The primary exteroceptive sensors include cameras, lidar, radar, and ultrasonic sensors.

Contributors: Yuxin Wang

W. Shi, Y. He, *Introduction to Autonomous Driving*,
https://doi.org/10.1007/978-3-031-99485-2_3

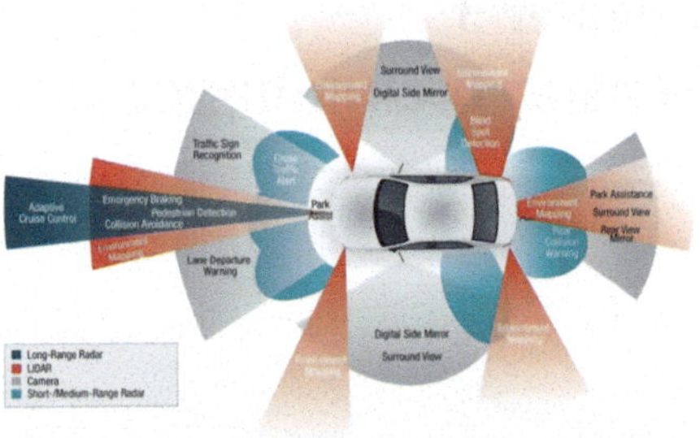

Fig. 3.1 Sensor usage in autonomous vehicles. (Image source: Machine Design)

Fig. 3.2 Example: KITTI image data[8]

3.2.1 Camera

Cameras function analogously to human vision by capturing detailed visual information. They are essential for object recognition, lane detection, and traffic sign interpretation. High-resolution imaging facilitates complex vision-based algorithms.

Cameras operate by capturing reflected light, which is focused through a lens onto an image sensor, typically a CCD or CMOS device. These sensors convert optical signals into electronic ones, forming digital images or video streams.

Many modern cameras offer high frame rates for real-time dynamic data capture. Key parameters such as lens configuration and sensor specifications influence the field of view, resolution, and depth perception. Cameras may be mounted at various points on the vehicle to ensure complete situational awareness (Fig. 3.2).

Camera output comprises digital frames (e.g., JPEG, PNG, or raw formats) with pixel values encoding light intensity, often in RGB channels. Some systems employ stereo vision or infrared/thermal modalities to enhance performance under challenging lighting.

Camera data is processed using edge detection, feature extraction, optical flow analysis, and deep learning methods to extract semantically meaningful information.

High-resolution imagery enables detection and classification of entities such as pedestrians, cyclists, vehicles, and road signage. Lane detection and tracking are also facilitated by camera systems, supporting lane-keeping and lane-changing maneuvers.

Camera systems are typically fused with radar, lidar, and ultrasonic data to enhance overall environmental awareness. However, they are sensitive to low light, precipitation, and glare, which underscores the necessity for sensor fusion.

Fig. 3.3 UD FinTech map created using LiDAR Hesai Pandar64

3.2.2 *Lidar*

Lidar (Light Detection and Ranging) measures distances using laser pulses to create detailed 3D representations of the environment, which is indispensable for obstacle detection and navigation in complex settings. By emitting rapid laser pulses and timing their return after object reflection, lidar calculates distances and constructs dense point clouds representing environmental geometry. The output is a point cloud comprising 3D coordinates (x, y, z) for each detected surface point. Figure 3.3 demonstrates a sample 3D map reconstructed via lidar.

Lidar is critical for generating precise maps and is particularly useful in low-visibility conditions where other sensors may fail.

3.2.3 *Radar*

Radar (Radio Detection and Ranging) utilizes radio waves to identify objects and estimate their distance and speed, offering robustness in adverse weather conditions such as fog, rain, or snow.

Radio waves emitted by the radar reflect off objects. By measuring time delays and Doppler shifts, radar systems determine object distances and relative velocities [6]. Based on operational range, radar is categorized as short-range (SRR), medium-range (MRR), or long-range (LRR) [18].

Radar outputs typically include point clouds encoding range, Doppler velocity, and azimuth, supporting applications like adaptive cruise control and collision avoidance.

3.2.4 Ultrasonic

Ultrasonic sensors detect nearby objects using high-frequency sound waves and are frequently employed in parking assistance and low-speed navigation.

Operating in the 23–40 kHz range, ultrasonic sensors emit sound chirps and measure the echo return time [20]. Each sensor combines a transmitter and receiver in a compact transducer.

These sensors output distance measurements from the vehicle's surface to nearby objects and are essential for obstacle detection in constrained environments.

3.3 Proprioceptive Sensors

Proprioceptive sensors provide information about the vehicle's own motion and spatial state. Two widely used sensors are GPS and IMU.

3.3.1 GPS (Global Positioning System)

GPS supplies geospatial data, enabling global positioning for route planning and tracking.

It triangulates satellite signals to compute position, commonly within meter-level accuracy. Differential GPS techniques enhance this accuracy.

GPS outputs include latitude, longitude, altitude, and velocity. These are often formatted according to the NMEA standard [10], using identifiers like GPGGA, GPGLL, and GPRMC.

Due to limitations in urban or obstructed areas, GPS is commonly augmented with IMU and lidar data for improved localization fidelity.

3.3.2 IMU (Inertial Measurement Unit)

IMUs integrate accelerometers and gyroscopes to measure linear and angular motion.

Their output includes tri-axial acceleration and rotational velocity, vital for maintaining accurate pose tracking, especially in GPS-denied environments.

IMUs are indispensable for continuous localization in tunnels or urban canyons. Combined with GPS, they ensure precise trajectory estimation. Figure 3.4 illustrates odometry from the KITTI dataset using GPS-IMU integration.

Autonomous vehicles rely on multiple sensors to perceive their surroundings accurately. However, raw sensor data is often noisy, misaligned, or subject to sys-

Fig. 3.4 Example: KITTI vehicle odometry

tematic errors. To ensure high-fidelity perception, sensor data must be preprocessed effectively. This section covers the principles and techniques used in sensor data calibration, cleaning, and fusion for autonomous driving.

3.4 Sensor Data Calibration

Autonomous vehicles rely on multiple sensors to perceive their surroundings accurately. However, raw sensor data is often noisy, misaligned, or subject to systematic errors. To ensure high-fidelity perception, sensor data must be preprocessed effectively. This section outlines the principles and techniques used in sensor data calibration, cleaning, and fusion for autonomous driving.

3.4.1 Sensor Data Calibration

Sensor calibration refers to the process of refining sensor measurements to ensure their accuracy and consistency. In autonomous vehicles, calibration is critical because any deviation in sensor output can lead to inaccurate perception and suboptimal decision-making.

Calibration is generally divided into two categories: intrinsic and extrinsic calibration [13].

Intrinsic Calibration Intrinsic calibration addresses the internal parameters of a sensor to achieve a geometrically accurate interpretation of its measurements. For cameras, this process—also known as camera re-sectioning—involves determining key characteristics such as focal length, principal point, and lens distortion coefficients. These intrinsic parameters are necessary for correcting image distortions so that the visual data corresponds precisely to real-world geometry.

A widely adopted method for intrinsic camera calibration employs a flat checkerboard pattern, as illustrated in Fig. 3.5. The known geometry of the checkerboard is compared with its projected appearance in captured images. By detecting corner points across multiple images taken from different perspectives, calibration algorithms compute intrinsic parameters such as the camera matrix and distortion coefficients. Accuracy is improved by capturing images from varied spatial

Fig. 3.5 Camera calibration using OpenCV (OpenCV: Camera Calibration, tutorials here). (Image source: Kaustubh Sadekar)

positions, ensuring comprehensive geometric coverage. The calibrated parameters enable the rectification of image distortions, thereby enhancing the reliability of perception tasks such as object recognition and depth estimation.

Extrinsic Calibration Extrinsic calibration defines the spatial relationships between different sensors, ensuring that their outputs are geometrically consistent. This is particularly vital in autonomous vehicles, where multimodal sensor integration—across cameras, LiDARs, and radars—is essential for accurate environmental modeling. Extrinsic calibration involves estimating the transformation parameters that map data from one sensor's coordinate frame into another's.

In systems combining a camera and a LiDAR, extrinsic calibration establishes the relative pose between the two. A typical method uses a calibration target that is visible to both sensors. Because LiDAR point clouds are sparser than image data, identifying corresponding features can be challenging. To overcome this, checkerboard targets embedded with reflective markers are often used to facilitate feature detection across both modalities. Recent developments propose 3D calibration targets with circular apertures, which improve feature localization in sparse point clouds. Once matching features are extracted from both sensors, the system computes a transformation matrix to align the LiDAR data with the camera frame.

Proper extrinsic calibration is essential for enabling accurate sensor fusion, which improves overall system robustness in the presence of occlusions, sensor malfunctions, or degraded environmental conditions. Advances in artificial intelligence and edge computing continue to enhance the speed and reliability of sensor fusion, enabling real-time performance in autonomous vehicle systems.

3.5 Sensor Data Cleaning

Raw sensor data obtained from autonomous vehicle sensors is rarely flawless. It is often affected by noise (random errors), outliers (anomalous readings), and missing data (incomplete observations). **Sensor data cleaning** refers to the process of addressing these issues to ensure that the vehicle's perception system receives reliable, high-quality input. Since different sensors—such as cameras and LiDAR—produce diverse data modalities, the corresponding cleaning techniques must be tailored accordingly. Clean sensor data is essential for enabling accurate obstacle detection, efficient path planning, and overall safe navigation.

3.5.1 Why Clean Sensor Data?

Sensors operate within complex and dynamic real-world environments. A camera may produce blurred images during rainfall, while a LiDAR system might capture erroneous reflections from wet or reflective surfaces. If such corrupted data is not appropriately processed, the perception system may misidentify non-existent objects or overlook genuine hazards. Cleaning procedures are designed to mitigate sensor-specific artifacts and deficiencies, thereby enhancing the robustness and accuracy of environmental perception.

3.5.2 Cleaning Camera Data

Camera sensors are responsible for capturing visual information that supports tasks such as object recognition, road sign interpretation, motion tracking, and general scene understanding. However, these sensors are susceptible to various distortions and sources of noise, particularly under conditions such as low illumination, high velocity, or inclement weather (e.g., fog or rain). These effects may manifest as pixel-level irregularities, color aberrations, graininess, or overall image blur.

A fundamental challenge in processing camera data is to suppress noise while preserving critical image features—such as edges, contours, and textures—that are essential for accurate object classification and semantic interpretation. The following subsection describes several advanced image denoising techniques and their practical implementation using Python and OpenCV.

3.5.3 State-of-the-Art Image Denoising Methods

- **Gaussian Blur**: Gaussian blur is a classical smoothing technique that reduces image noise by averaging pixel intensities within a local neighborhood. Although effective at noise suppression, this method may blur salient features such as edges, thereby degrading performance in downstream detection tasks [7].
- **Bilateral Filter**: The bilateral filter performs edge-preserving denoising by considering both spatial proximity and intensity similarity when computing weighted averages. This dual-domain filtering helps reduce noise while maintaining edge integrity, making it suitable for applications requiring high-fidelity detail preservation, including medical imaging and high-dynamic-range processing [24].
- **Median Filtering**: Median filtering is a non-linear approach that substitutes each pixel with the median value of its local neighborhood [9]. It is especially effective against salt-and-pepper noise and preserves edge structures more reliably than linear methods such as Gaussian blur.
- **Non-Local Means Denoising**: Non-local means (NLM) denoising leverages image self-similarity by comparing pixel patches across the entire image. This method averages similar patches regardless of spatial proximity, achieving strong noise reduction while preserving fine-scale image features and textures.
- **Deep Learning-Based Denoising**: Deep learning models, particularly those based on Convolutional Neural Networks (CNNs), have demonstrated exceptional denoising performance [14]. Architectures such as DnCNN (Denoising Convolutional Neural Network) are trained to distinguish noise patterns from structural image content using large datasets. These methods offer state-of-the-art results in low-quality or highly variable imaging conditions [23], albeit with greater computational demands.
- **Wavelet Transform Denoising**: Wavelet-based methods decompose an image into multiple frequency subbands, where noise is predominantly found in the high-frequency components. By applying thresholding in the wavelet domain and reconstructing the image, it is possible to suppress noise effectively while retaining structural features such as edges and textures.

3.5.4 Example: Denoising an Image Using OpenCV in Python

Here, two widely used denoising techniques are implemented using Python and OpenCV: **Bilateral Blur** and **Non-Local Means Denoising**. Gaussian noise is artificially added to demonstrate the effectiveness of each method.

```
import numpy as np
import cv2
from skimage import data, img_as_float
import matplotlib.pyplot as plt
```

```
# Load an example image (astronaut) from skimage
image = data.astronaut()

# Display the original image
plt.figure(figsize=(15, 5))
plt.subplot(1, 4, 1)
plt.title("Original␣Image")
plt.imshow(image)
plt.axis('off')

# Add Gaussian noise with zero mean and standard deviation of 25
mean = 0
stddev = 25
gaussian_noise = np.random.normal(mean, stddev, image.shape)

# Add noise to image and ensure pixel values remain in valid range [0, 255]
noisy_image = np.clip(image + gaussian_noise, 0, 255).astype(np.uint8)

# Display the noisy image
plt.subplot(1, 4, 2)
plt.imshow(noisy_image)
plt.title('Noisy␣Image')
plt.axis('off')

# Apply Bilateral Filter with specific diameter and sigma values
bilateral_blurred = cv2.bilateralFilter(noisy_image, 9, 75, 75)

# Display the result of Bilateral Filter
plt.subplot(1, 4, 3)
plt.imshow(bilateral_blurred)
plt.title('Bilateral␣Blurred')
plt.axis('off')

# Apply Non-Local Means Denoising with tuned parameters
nlm_denoised = cv2.fastNlMeansDenoisingColored(noisy_image, None, 13, 10, 7,
      21)

# Display the result of Non-Local Means Denoising
plt.subplot(1, 4, 4)
plt.imshow(nlm_denoised)
plt.title('NLM␣Denoised')
plt.axis('off')

# Adjust subplot layout and show all plots
plt.tight_layout()
plt.show()

# Compute PSNR (Peak Signal-to-Noise Ratio) to quantify quality improvement
psnr_bilateral = cv2.PSNR(image, bilateral_blurred)
psnr_nlm = cv2.PSNR(image, nlm_denoised)
print(f"PSNR␣(Bilateral␣Blur)␣=␣{psnr_bilateral:.2f}␣dB")
print(f"PSNR␣(NLM␣Denoised)␣=␣{psnr_nlm:.2f}␣dB")
```

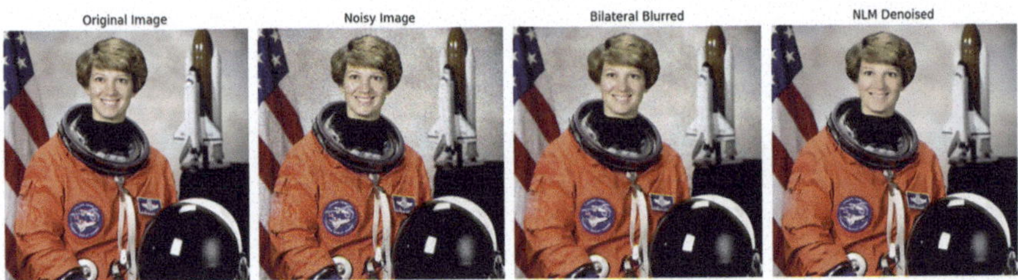

Fig. 3.6 Results of OpenCV image denoising

After executing the script, the visual and quantitative results are displayed as shown in Fig. 3.6.

3.5.5 Cleaning LiDAR Data

LiDAR sensors produce 3D point clouds that often require preprocessing to remove noise, eliminate outliers, and reduce data density for efficient downstream processing. The Point Cloud Library (PCL) offers a range of robust filtering techniques categorized into three principal areas: Noise Reduction, Outlier Removal, and Point Cloud Downsampling.

1. **Noise Reduction**: LiDAR noise can result from sensor inaccuracies, environmental interference, and dynamic scene changes. The following methods effectively smooth noisy measurements:

   - **pcl::BilateralFilter**: Weights points based on both spatial proximity and intensity similarity to preserve sharp edges while reducing high-frequency noise.
   - **pcl::MedianFilter**: A non-linear filter that substitutes each point with the median of its neighbors, effectively removing impulsive noise while retaining structure.
   - **pcl::filters::GaussianKernel**: Applies Gaussian smoothing by weighting points based on their distance to neighbors, thereby attenuating fine-grained noise.

2. **Outlier Removal**: Outliers are isolated measurements often caused by reflective surfaces or dynamic elements. These methods identify and exclude such points:

   - **pcl::StatisticalOutlierRemoval (SOR)**: Calculates the mean and standard deviation of a point's neighbors and removes those that deviate significantly beyond a threshold.
   - **pcl::RadiusOutlierRemoval (ROR)**: Discards points with insufficient neighboring points within a defined radius, targeting sparsely distributed noise.

- **pcl::ModelOutlierRemoval**: Filters points based on their deviation from a pre-defined geometric model, suitable for structured environments like urban roads.

3. **Point Cloud Downsampling**: High-resolution LiDAR captures large volumes of data that must be reduced for computational efficiency while retaining essential spatial features:

- **pcl::VoxelGrid**: Segments space into uniform voxels and replaces all points in each voxel with a representative point, often the centroid.
- **pcl::FarthestPointSampling (FPS)**: Selects points iteratively to maximize spatial distribution, maintaining structural representativeness.
- **pcl::UniformSampling**: Uses a regular grid to sample points evenly across the cloud, ensuring uniform spatial coverage.

3.5.6 Example: Filter Point Cloud Data Using Open3D in Python

This example demonstrates two common filtering techniques using Open3D: **Voxel Grid Filter** for downsampling and **Statistical Outlier Removal** for denoising. The input dataset is sourced from the Open3D built-in point cloud collection (https://www.open3d.org/docs/latest/tutorial/data/index.html).

```
import open3d as o3d
import numpy as np

# Load an example point cloud from the Open3D Redwood dataset
dataset = o3d.data.PCDPointCloud()

# Read the point cloud file
print("Load a PCD point cloud and visualize it")
pcd = o3d.io.read_point_cloud(dataset.path)

# Visualize the original point cloud
o3d.visualization.draw_geometries([pcd],
                                  zoom=0.3412,
                                  front=[0.4257, -0.2125, -0.8795],
                                  lookat=[2.6172, 2.0475, 1.532],
                                  up=[-0.0694, -0.9768, 0.2024])

# Apply voxel grid downsampling with voxel size of 0.02 meters
print("Downsample the point cloud with a voxel size of 0.02")
voxel_down_pcd = pcd.voxel_down_sample(voxel_size=0.02)

# Visualize the downsampled point cloud
o3d.visualization.draw_geometries([voxel_down_pcd],
                                  zoom=0.3412,
```

```
                                    front=[0.4257, -0.2125, -0.8795],
                                    lookat=[2.6172, 2.0475, 1.532],
                                    up=[-0.0694, -0.9768, 0.2024])

# Define function to separate and visualize inliers and outliers
def display_inlier_outlier(cloud, ind):
    # Select inlier points by index
    inlier_cloud = cloud.select_by_index(ind)
    # Select outlier points by inverting the index
    outlier_cloud = cloud.select_by_index(ind, invert=True)

    print("Showing␣outliers␣(red)␣and␣inliers␣(gray):")
    # Color-code the points for visualization
    outlier_cloud.paint_uniform_color([1, 0, 0]) # Red for outliers
    inlier_cloud.paint_uniform_color([0.8, 0.8, 0.8]) # Gray for inliers

    # Display inliers and outliers side-by-side
    o3d.visualization.draw_geometries([inlier_cloud, outlier_cloud],
                                      zoom=0.3412,
                                      front=[0.4257, -0.2125, -0.8795],
                                      lookat=[2.6172, 2.0475, 1.532],
                                      up=[-0.0694, -0.9768, 0.2024])

# Apply statistical outlier removal filter
print("Statistical␣outlier␣removal")
cl, ind = voxel_down_pcd.remove_statistical_outlier(nb_neighbors=20,
                                                    std_ratio=2.0)
# Visualize filtered result with outliers highlighted
display_inlier_outlier(voxel_down_pcd, ind)
```

After executing the script, the visual results will appear as shown in Fig. 3.7.

(a) Original point cloud. **(b)** Point cloud after voxel grid filter. **(c)** Point cloud after SOR (showing outliers in red and inliers in gray).

Fig. 3.7 Results of Open3D point cloud filtering. (**a**) Original point cloud. (**b**) Point cloud after voxel grid filter. (**c**) Point cloud after SOR (showing outliers in red and inliers in gray)

3.6 Sensor Data Fusion

Multi-sensor data fusion, also referred to as data fusion, information fusion, or sensor fusion, is a foundational component of autonomous driving. It involves combining, merging, or integrating homogeneous or heterogeneous data from multiple sensors to achieve a more accurate, robust, and comprehensive understanding of the environment. The primary objectives of sensor fusion are to enhance perception, improve localization, and infer critical information that may not be discernible from any single sensor alone [19].

Sensor fusion is inherently multidisciplinary, requiring expertise in signal processing, machine learning, statistics, artificial intelligence, and robotics. The process typically includes sensor calibration, synchronization, noise filtering, data alignment, and high-level semantic interpretation. Given the diversity of sensors in autonomous vehicles—including cameras, LiDAR, radar, IMUs (Inertial Measurement Units), GNSS (Global Navigation Satellite Systems), and V2X (Vehicle-to-Everything) communication systems—developing reliable and computationally efficient fusion strategies is essential for enabling safe and intelligent navigation.

3.6.1 Sensor Fusion in Perception Systems

Perception systems in autonomous vehicles rely on heterogeneous sensors, each with distinct advantages and limitations. Cameras capture high-resolution, color-rich imagery suitable for object classification, yet their performance degrades under low-light or adverse weather conditions. LiDAR produces accurate 3D point clouds, effective for shape and distance estimation, but is sensitive to environmental obstructions such as rain or fog. Radar excels in long-range detection and maintains functionality under adverse weather, though it has lower spatial resolution. Ultrasonic sensors are reliable for short-range detection and low-speed maneuvering but lack broader scene awareness.

By integrating information from these diverse sensors, sensor fusion mitigates individual sensor weaknesses and enables robust perception. For instance, the fusion of radar and camera data allows simultaneous extraction of visual details and motion cues, enhancing dynamic obstacle detection. Similarly, combining LiDAR's spatial precision with camera imagery supports both geometric understanding and semantic labeling. These integrated perception systems allow vehicles to detect, classify, and track objects reliably, even in complex or degraded sensing environments [21, 5].

3.6.2 Sensor Fusion in Localization Systems

Accurate localization is critical for autonomous navigation. While GNSS provides global positioning, it can suffer from signal degradation in environments such as tunnels or urban canyons. IMUs offer high-frequency motion updates but accumulate error over time (drift). Sensor fusion approaches compensate for such limitations by combining complementary data sources.

The integration of GPS, IMU, and wheel encoder data improves localization accuracy and reliability. Algorithms such as Kalman filters and particle filters are used to optimally fuse noisy and partial data from these sources, yielding stable and continuous pose estimates. Such fusion strategies enable autonomous vehicles to localize themselves effectively, even when one or more sensors are compromised [21, 5].

3.6.3 Sensor Fusion Approaches

Sensor fusion strategies in autonomous systems are generally categorized into three levels: high-level fusion (HLF), low-level fusion (LLF), and mid-level fusion (MLF) [1].

High-Level Fusion (HLF) In HLF, each sensor independently performs detection or tracking, and the outputs are then combined. For example, radar and LiDAR data are processed separately, and their results are fused to reinforce object hypotheses. This method reduces computational demands and simplifies data alignment. However, it may discard fine-grained information, potentially leading to lower confidence in ambiguous scenarios.

Low-Level Fusion (LLF) LLF integrates raw sensor data directly, preserving all available information. This comprehensive input improves detection accuracy and robustness. For instance, raw LiDAR and camera data may be jointly processed in a multi-stage detection pipeline. LLF, however, requires meticulous sensor calibration and synchronization, as well as significant computational resources to manage high-dimensional data and compensate for vehicle ego-motion.

Mid-Level Fusion (MLF) MLF represents a compromise between HLF and LLF. In this approach, features (e.g., object position, velocity, appearance) are extracted from each sensor stream and then fused. MLF enables more flexible integration than HLF, allowing the use of only salient features, but it may still lose important contextual information contained in raw data.

3.6.4 Sensor Fusion Techniques and Algorithms

Sensor fusion methods fall into two primary categories: classical algorithms and deep learning-based techniques.

Classical Sensor Fusion Methods Classical techniques rely on probabilistic and model-based approaches to integrate data. Kalman filters and particle filters are widely used to combine measurements with known uncertainty characteristics, providing statistically optimal state estimates for vehicle localization [2]. Additionally, fuzzy logic systems have been used in decision-making for structured environments, such as traffic intersections [17].

Deep Learning-Based Sensor Fusion Deep learning offers a data-driven approach to fusion. CNNs [14], RNNs [22], and transformer architectures are capable of learning hierarchical features from raw sensor data. Algorithms like YOLO [16] use real-time object detection pipelines that can incorporate both image and LiDAR modalities. SSD [12] and CenterNet [4] further enhance 2D and 3D object detection.

Advanced 3D fusion architectures include VoxelNet [25] and PointNet [15]. VoxelNet partitions point cloud data into voxel grids and processes them using 3D convolutional layers, integrating these with features from camera images for enhanced detection. PointNet directly processes raw point cloud data without voxelization, enabling fine-grained recognition even in sparse or unstructured environments. These approaches are fundamental to achieving robust and scalable perception in autonomous systems.

3.7 Challenges in Data Processing

Real-time sensor data processing presents several challenges. Cameras and LiDARs generate high-frequency data streams—typically around 30 frames per second for cameras and 10 Hz for LiDAR—requiring algorithms to process information with minimal latency. However, excessive filtering can inadvertently remove critical scene elements, such as slender poles or faint traffic signs. Engineers must therefore fine-tune preprocessing parameters using diverse driving datasets to balance efficiency and accuracy.

Environmental variability further complicates sensor data cleaning. Rain, fog, and direct sunlight can introduce artifacts, leading to degraded sensor performance. For example, LiDAR beams may scatter in heavy precipitation, producing false returns, while camera images can be overexposed by glare. Adaptive filtering techniques provide some resilience, but they must be dynamically adjusted to changing conditions.

Sensor misalignment poses another challenge in multi-sensor systems. Even minor calibration errors can lead to spatial inconsistencies in fused data. For instance, if camera and LiDAR sensors are not precisely aligned, the same pedestrian may appear at different locations in the reconstructed scene. While automated calibration techniques mitigate this issue, achieving sub-centimeter precision in real-time, dynamic environments remains difficult.

Outlier detection is also critical. Sensors may report spurious "phantom" objects—such as reflections from glass or unintended radar echoes. Simple threshold-based approaches can remove obvious artifacts, but more robust discrimination requires machine learning-based outlier classifiers capable of distinguishing noise from true environmental features.

Missing data presents a further challenge. Temporary occlusions, such as other vehicles obstructing key landmarks, can result in perception gaps. Predictive methods, including interpolation and AI-based reconstruction, aim to fill these gaps. However, they carry the risk of propagating inaccuracies if the underlying assumptions are violated.

Lastly, computational constraints limit real-time processing. Autonomous systems must handle massive sensor data volumes within strict latency budgets. Engineers employ parallel processing and hardware acceleration—such as GPUs and AI accelerators—to maintain throughput. Designing systems that balance computational load with perceptual fidelity remains an ongoing area of research and optimization.

3.8 Datasets

Before introducing specific datasets used in autonomous driving, it is essential to understand the range of tasks and data modalities involved. Autonomous driving represents a highly complex and multidisciplinary domain, encompassing tasks such as perception, prediction, planning, control, and the increasingly prominent end-to-end driving models. Each of these components relies on different types of data and annotation standards.

Datasets in this field are often designed to address particular subtasks and are collected using a wide array of sensing platforms. Common data sources include onboard sensors such as cameras, LiDARs, radars, and GPS/IMU units, as well as external inputs like vehicle-to-everything (V2X) communications, aerial drone footage, and infrastructure-based sensors. This diversity accommodates modeling of varied driving environments, interaction contexts, and sensor configurations.

3.8.1 Task-Oriented Datasets

Datasets form the foundation of model training, evaluation, and benchmarking. Within the autonomous driving pipeline, raw sensor data is processed through a series of core tasks. Below, the major categories are outlined along with their associated input modalities and typical dataset characteristics:

- **Perception and Localization:** Tasks include 2D/3D object detection, semantic segmentation, object tracking, and simultaneous localization and mapping (SLAM).
  Input: Camera images, LiDAR point clouds, and radar signals.
  Use: These tasks enable interpretation of the vehicle's immediate environment—identifying vehicles, pedestrians, lanes, and infrastructure. Tracking and SLAM require temporally dense, time-stamped annotations to model motion and vehicle pose.
- **Prediction:** Forecasting the future behaviors or trajectories of dynamic agents such as vehicles, pedestrians, and cyclists.
  Input: Temporal sensor sequences, object trajectories, velocity vectors, and semantic maps.
  Use: Supports anticipatory decision-making by modeling the expected evolution of the environment.
- **Planning and Control:** Tasks include path planning, behavior planning, and motion control.
  Input: High-level route descriptions, vehicle state data, dynamic object predictions, and HD maps.
  Use: These components translate abstract driving goals into executable trajectories and control signals.
- **End-to-End Driving Policy Learning:** Methods that aim to directly map raw sensor inputs to driving actions, bypassing traditional modular pipelines.
  Input: Typically camera data (optionally multimodal), aligned with control commands (steering, acceleration, braking).
  Use: Requires large-scale annotated datasets from real-world or simulated driving for supervised or reinforcement learning.

Different tasks necessitate different data structures. Frame-based annotations suffice for detection, whereas prediction and tracking demand sequential, densely annotated data. The selection of a dataset must align with task objectives, sensor modalities, and deployment context.

3.8.2 Popular Datasets in Autonomous Driving

This section presents three widely adopted datasets that serve as benchmarks for algorithm development and evaluation in autonomous driving.

KITTI Dataset

The KITTI Vision Benchmark Suite,[1] developed by the Karlsruhe Institute of Technology and Toyota Technological Institute at Chicago, is one of the pioneering datasets in this domain. It features data captured from a vehicle outfitted with high-resolution color and grayscale cameras, a Velodyne LiDAR scanner, and a GPS/IMU system. Scenes include urban roads, rural areas, and highways. Core tasks include stereo vision, optical flow, visual odometry, 3D object detection, and 3D tracking.

nuScenes Dataset

nuScenes,[2] created by Motional, is a comprehensive dataset supporting multiple autonomous driving tasks. Key components include:

- **nuPlan:** A planning benchmark with 1200 hours of driving data across Boston, Pittsburgh, Las Vegas, and Singapore. Sensor data is released for 120 hours, including 5 LiDARs, 8 cameras, GPS, and IMU.
- **nuScenes:** A dataset comprising 1000 scenes (each 20 seconds), annotated for 3D detection, tracking, prediction, and segmentation. It includes 6 cameras, 5 radars, 1 LiDAR, GPS, and IMU.
- **nuImages:** A large-scale dataset for 2D vision tasks. It contains 93,000 annotated and 1.1 million unannotated images from 150 hours of driving in Boston and Singapore.

Waymo Open Dataset

The Waymo Open Dataset[3] offers an industry-grade sensor suite and extensive annotations. Highlights include:

- **Perception Dataset:** High-resolution data with 2030 annotated driving segments.
- **Motion Dataset:** Agent trajectories and corresponding maps for 103,354 segments.
- **End-to-End Driving Dataset:** Contains camera imagery with 360-degree coverage and routing metadata for 5000 segments.

[1] The KITTI Vision Benchmark Suite: https://www.cvlibs.net/datasets/kitti/.

[2] nuScenes Dataset: https://www.nuscenes.org/.

[3] Waymo Open Dataset: https://waymo.com/open/.

3.8.3 Conclusion and Future Directions

As autonomous driving matures, future datasets will increasingly capture rare events, long-tail phenomena, and edge-case scenarios. This shift reflects a need for real-world robustness beyond standard benchmarks. Simulation-based datasets are also gaining traction, particularly in tasks where annotation is cost-prohibitive.

Looking ahead, dataset development will emphasize diversity in geographic, environmental, and behavioral dimensions, high-fidelity sensor replication, and sim-to-real generalization to support the transition from research prototypes to production-grade autonomous systems.

3.9 Hands-on Exercises

3.9.1 A Comparison of Real and Simulated Sensor Data

Objective In this exercise, students will adjust CARLA's camera and LiDAR sensor parameters within the BlueICE framework to make simulated outputs visually and structurally resemble real-world sensor data collected from the ICAT testbed.

Task Using the same trajectory in both simulation and the real-world dataset, students will iteratively tune parameters such as camera exposure, FOV, resolution, and LiDAR field of view and noise models. The goal is to minimize perceptual and structural differences between simulated and real sensor outputs.

Deliverable Submit a pair of side-by-side comparisons (image and point cloud), along with a short written summary describing your tuning strategy and the most noticeable differences. Be prepared to present your results and test whether others can distinguish between real and simulated data.

3.9.2 IMU-GNSS Sensor Fusion on the KITTI Dataset (Advanced)

Objective Understand how IMU and GNSS data can be fused to estimate vehicle pose using an Unscented Kalman Filter.

Task Explore the UKFM implementation of IMU-GNSS fusion on the KITTI dataset. Study how the estimator tracks position and orientation over time. Modify parameters or sensor noise models and observe their effects on trajectory accuracy.

Deliverable Submit a plot comparing ground truth and estimated trajectories, along with a brief explanation of what you changed and how it affected performance.[4]

3.9.3 BEVFusion: A Simple and Robust LiDAR-Camera Fusion Framework (Optional)

Objective Gain experience in fusing camera and LiDAR data for object detection using a deep learning-based method.

Task Clone and set up the BEVFusion repository.[5] Use the provided scripts to run inference on the sample dataset. Analyze how BEVFusion integrates multi-modal inputs—specifically, image and LiDAR data—for object detection in the bird's-eye view (BEV). Optionally, experiment with disabling one modality (camera or LiDAR) to evaluate their individual contributions.

Deliverable Submit (1) a screenshot of the 3D detection result visualized in BEV format, and (2) a brief reflection (150 words) comparing the roles of camera and LiDAR in detection accuracy and spatial localization.

Recommended Papers to Read

1. Yeong, D. J., Velasco-Hernandez, G., Barry, J., & Walsh, J. (2021). Sensor and sensor fusion technology in autonomous vehicles: A review. Sensors, 21(6), 2140. [21]
2. Liu, M., Yurtsever, E., Fossaert, J., Zhou, X., Zimmer, W., Cui, Y., ... & Knoll, A. C. (2024). A survey on autonomous driving datasets: Statistics, annotation quality, and a future outlook. IEEE Transactions on Intelligent Vehicles. [11]

References

1. Banerjee, K., et al. (2018). Online camera lidar fusion and object detection on hybrid data for autonomous driving. In *2018 IEEE Intelligent Vehicles Symposium (IV)* (pp. 1632–1638). IEEE.
2. Barrau, A., & Bonnabel, S. (2018). Invariant Kalman filtering. *Annual Review of Control, Robotics, and Autonomous Systems*, *1*(1), 237–257.
3. Campbell, S., et al. (2018). Sensor technology in autonomous vehicles: A review. In *2018 29th Irish Signals and Systems Conference (ISSC)* (pp. 1–4). IEEE.
4. Duan, K., et al. (2019). Centernet: Keypoint triplets for object detection. In *Proceedings of the IEEE/CVF International Conference on Computer Vision* (pp. 6569–6578).

[4] See the original implementation: https://caor-mines-paristech.github.io/ukfm/auto_examples/imugnss.html.

[5] See the original implementation: https://github.com/ADLab-AutoDrive/BEVFusion.

5. Fayyad, J., et al. (2020). Deep learning sensor fusion for autonomous vehicle perception and localization: A review. *Sensors, 20*(15), 4220.
6. Gamba, J. (2020). *Radar signal processing for autonomous driving* (Vol. 1456). Springer.
7. Gedraite, E. S., & Hadad, M. (2011). Investigation on the effect of a Gaussian Blur in image filtering and segmentation. In *Proceedings ELMAR-2011* (pp. 393–396). IEEE.
8. Geiger, A. Lenz, P., & Urtasun, R. (2012). Are we ready for autonomous driving? The KITTI vision benchmark suite. In *Conference on Computer Vision and Pattern Recognition (CVPR)*.
9. Hwang, H., & Haddad, R. A. (1995). Adaptive median filters: New algorithms and results. *IEEE Transactions on Image Processing, 4*(4), 499–502.
10. Langley, R. B. (1995). NMEA 0183: A GPS receiver interface standard. *GPS World, 6*(7), 54–57. https://gge.ext.unb.ca/Resources/gpsworld.july95.pdf
11. Liu, M., Yurtsever, E., Fossaert, J., Zhou, X., Zimmer, W., Cui, Y., Zagar, B. L. & Knoll, A. C. (2024). A survey on autonomous driving datasets: Statistics, annotation quality, and a future outlook. IEEE Transactions on Intelligent Vehicles.
12. Liu, W., et al. (2016). SSD: Single shot multibox detector. In *Computer Vision–ECCV 2016: 14th European Conference, Amsterdam, The Netherlands, October 11–14, 2016, Proceedings, Part I 14* (pp. 21–37). Springer.
13. Mirzaei, F. M. (2013). *Extrinsic and intrinsic sensor calibration*. PhD thesis. University of Minnesota.
14. O'shea, K., & Nash, R. (2015). *An introduction to convolutional neural networks*. Preprint. arXiv:1511.08458.
15. Qi, C. R., et al. (2017). PointNet: Deep learning on point sets for 3D classification and segmentation. In *Proceedings of the IEEE Conference on Computer Vision and Pattern Recognition* (pp. 652–660).
16. Redmon, J., et al. (2016). You only look once: Unified, real-time object detection. In *Proceedings of the IEEE Conference on Computer Vision and Pattern Recognition* (pp. 779–788).
17. Uzunsoy, E. (2018). A brief review on fuzzy logic used in vehicle dynamics control. *Journal of Innovative Science and Engineering (JISE), 2*(1), 1–7.
18. Vargas, J., et al. (2021). An overview of autonomous vehicles sensors and their vulnerability to weather conditions. *Sensors, 21*(16), 5397.
19. Velasco-Hernandez, G., et al. (2020). Autonomous driving architectures, perception and data fusion: A review". In *2020 IEEE 16th International Conference on Intelligent Computer Communication and Processing (ICCP)* (pp. 315–321). https://doi.org/10.1109/ICCP51029.2020.9266268
20. Xu, W., et al. (2018). Analyzing and enhancing the security of ultrasonic sensors for autonomous vehicles. *IEEE Internet of Things Journal, 5*(6), 5015–5029. https://doi.org/10.1109/JIOT.2018.2867917
21. Yeong, D. J., et al. (2021). Sensor and sensor fusion technology in autonomous vehicles: A review. *Sensors, 21*(6), 2140.
22. Yu, Y., et al. (2019). A review of recurrent neural networks: LSTM cells and network architectures. *Neural Computation, 31*(7), 1235–1270.
23. Zhang, K., et al. (2017). Beyond a gaussian denoiser: Residual learning of deep CNN for image denoising. *IEEE Transactions on Image Processing, 26*(7), 3142–3155.
24. Zhang, M., & Gunturk, B. K. (2008). Multiresolution bilateral filtering for image denoising. *IEEE Transactions on Image Processing, 17*(12), 2324–2333.
25. Zhou, Y., & Tuzel, O. (2018). Voxelnet: End-to-end learning for point cloud based 3d object detection. In *Proceedings of the IEEE Conference on Computer Vision and Pattern Recognition* (pp. 4490–4499).

Chapter 4
V2X Communications

4.1 Introduction

The transformation of transportation through autonomous driving is not solely a matter of more intelligent vehicles but of increasingly connected environments. Even with advanced perception systems, autonomous vehicles (AVs) face limitations arising from occlusions, restricted sensor ranges, or delayed detection of environmental changes. Such constraints inhibit real-time, fully informed decision-making, particularly in complex urban scenarios. Vehicle-to-Everything (V2X) communication addresses these challenges by allowing vehicles to transmit and receive information beyond the scope of their onboard sensors.

The foundational role of V2X communication technologies is critical in enabling safe and efficient autonomous driving. Kutila et al. offer a comprehensive overview of how Cellular V2X (C-V2X) facilitates automated driving by establishing vehicular communication in diverse contexts, emphasizing low-latency and high-reliability links for safety-critical maneuvers [9]. Similarly, Hobert et al. examine enhancements in V2X communication protocols that support cooperative autonomous driving, detailing developments in standardization and integration with intelligent transportation systems [7]. Bagheri et al. outline the technical roadmap for 5G NR-V2X, highlighting how fifth-generation wireless technologies improve inter-vehicle connectivity and cooperation [1]. Cinque et al. contribute by identifying the service requirements for connected and autonomous driving and aligning these with future communication technologies [2].

Through direct Vehicle-to-Vehicle (V2V) communication, AVs can access information about positions, speeds, acceleration, braking status, and intended maneuvers of surrounding vehicles, including those obscured by large vehicles, road curvature, or urban infrastructure. This capability enables proactive responses to dynamic changes, such as sudden braking or merging. Vehicle-to-Infrastructure (V2I) communication allows AVs to interact with smart infrastructure, including traffic signals and road signage systems. These systems convey data such as signal

W. Shi, Y. He, *Introduction to Autonomous Driving*,
https://doi.org/10.1007/978-3-031-99485-2_4

phase and timing (SPaT), temporary speed limits, construction updates, and lane closures. Such information aids in trajectory optimization, idle time reduction, and safer decision-making at intersections and merging points.

Vehicle-to-Pedestrian (V2P) communication extends situational awareness to pedestrians and cyclists equipped with connected devices. For example, an AV can receive a warning about an imminent pedestrian crossing, even if that pedestrian is hidden from direct view. Cyclists transmitting their location and velocity data also benefit from enhanced detection in shared road spaces.

Vehicle-to-Network (V2N) communication links AVs with cloud services and remote data aggregators. These systems provide high-definition map updates, predictive traffic analytics, long-range hazard alerts, and emergency routing information. When combined, these V2X modalities enable AVs to transcend reactive, sensor-only perception, moving toward a cooperative, informed model where decisions incorporate broader contextual and anticipatory data.

Integrating these multilayered external inputs, V2X communication strengthens AV decision-making with earlier and more accurate predictions and maneuvers. Thus, V2X does not merely extend perception; it cultivates a cooperative and anticipatory environment essential for next-generation autonomous systems.

V2X facilitates communication not only among nearby vehicles (V2V) but also with infrastructure (V2I), pedestrian devices (V2P), and the cloud (V2N). Real-time updates concerning road closures, construction zones, signal timing, and dynamic speed limits significantly improve the situational accuracy of AV decisions. Applications such as cooperative adaptive cruise control, platooning, and intersection coordination rely fundamentally on such exchanges. Therefore, V2X stands as a cornerstone for the safe and efficient operation of autonomous systems in mixed traffic environments.

A primary obstacle to the widespread deployment of V2X lies in the absence of a standardized, reliable communication infrastructure capable of supporting low-latency, high-throughput, and secure data exchange. This chapter explores the key technologies underpinning V2X, particularly Dedicated Short-Range Communication (DSRC) and Cellular V2X (C-V2X), while addressing their practical applications and the technical challenges that must be overcome to embed V2X within intelligent transportation systems.

V2X communication redefines transportation through cooperative awareness and coordinated vehicular behavior. It enables the sharing of planned trajectories, prediction of upcoming road conditions, and preemptive responses to hazards or inefficiencies. Essential services, including emergency vehicle prioritization, adaptive traffic signal negotiation, cooperative lane merging, and dynamic speed harmonization, depend on V2X to scale autonomous driving beyond isolated capabilities and into systems-level intelligence.

One of the most significant barriers to the realization of V2X's full potential is the lack of ubiquitous infrastructure with the necessary low latency and high reliability to support large-scale, heterogeneous communication. Additionally, interoperability across manufacturers and regulatory consensus are vital for seamless integration.

As a result, V2X emerges not only as a technical innovation but as a sociotechnical challenge fundamental to the evolution of autonomous mobility.

Despite its clear advantages, the broad adoption of V2X faces ongoing challenges, including latency management, cybersecurity vulnerabilities, and the necessity for substantial infrastructure investment. This chapter presents a detailed examination of V2X protocols such as DSRC and C-V2X, practical deployments, and emerging solutions like OpenIntersection, a software-defined architecture designed to optimize intersection coordination.

4.2 OpenIntersection and Edge Computing

As discussed in the earlier sections of this chapter, Vehicle-to-Everything (V2X) communication provides autonomous vehicles with a broader understanding of their environment by enabling information exchange with infrastructure, pedestrians, and other vehicles. However, while protocol-level innovations such as DSRC and C-V2X improve communication reliability and bandwidth, they do not fundamentally alter the static and opaque nature of current traffic infrastructure. Traditional intersections, even when connected, function as passive broadcast systems, providing limited, non-adaptive data such as signal phase and timing (SPaT). These systems lack the semantic flexibility and contextual responsiveness needed to support the highly dynamic interactions required by autonomous vehicles in real-world urban settings.

OpenIntersection is introduced to address these limitations. It reimagines the traffic intersection as a programmable, edge-assisted, software-defined entity that can actively coordinate with connected vehicles. Instead of treating infrastructure as a fixed-function component, OpenIntersection treats it as a semantic interface: an intelligent system capable of interpreting intent, negotiating priorities, and managing multi-agent interactions through policy-driven APIs.

The motivation for OpenIntersection arises from several critical deficiencies in current V2X-enabled infrastructure:

- Lack of semantic interoperability: Existing RSUs broadcast messages without the ability to interpret or reason about the intentions of individual vehicles. OpenIntersection introduces a shared abstraction layer where heterogeneous agents can interact through standardized semantic primitives, regardless of their underlying communication stacks. This semantic abstraction supports both human-interpretable and machine-readable formats, enabling high-level coordination strategies without requiring every agent to operate on the same low-level protocol.
- Static and non-programmable behavior: Current systems operate on rigid scheduling cycles and pre-configured rules. OpenIntersection supports dynamic, on-the-fly policy updates and behavior adaptation. For example, it can prioritize public transit or emergency vehicles based on contextual data, or dynamically

reallocate traversal slots to minimize congestion. Through its programmable API, cities can encode real-time priorities that reflect not just traffic volumes, but societal values such as equity, accessibility, and sustainability.
- Insufficient responsiveness: Relying on cloud-based orchestration introduces latency that is unacceptable for real-time decision-making. By deploying computation at the edge, OpenIntersection ensures low-latency negotiation and control, supporting time-sensitive V2X services like cooperative lane merging or intersection crossing. Edge processing also enhances system robustness, allowing local control logic to persist even during intermittent connectivity or cloud failures.
- Fragmented protocol ecosystems: Vehicles using different V2X technologies (e.g., DSRC vs. C-V2X) often cannot cooperate meaningfully. OpenIntersection mitigates this by acting as a semantic intermediary, translating between protocol-specific encodings and exposing a unified interface for coordination. This ensures that vehicles with varying capabilities and vendors can participate in shared traffic management processes without needing to converge on a single communication stack.
- The OpenIntersection framework introduces hybrid communication models that go beyond the fixed-function broadcasting of conventional infrastructure. Drawing from the OpenIntersection architecture proposed by He and Shi[5], this model enables infrastructure to not only disseminate general-purpose traffic updates via push-based channels but also to serve as a queryable resource through pull-based interaction. For instance, infrastructure may continue to broadcast SPaT messages and occupancy signals, while simultaneously exposing programmable APIs through which vehicles can submit high-level semantic queries or declare navigational intent (e.g., "I intend to turn left in 3.2 seconds").

This duality allows OpenIntersection to support differentiated levels of service: time-critical alerts can be pushed proactively, while context-sensitive coordination (e.g., negotiation of intersection priority or retrieval of intersection-specific policies) can be handled through explicit pull mechanisms. Furthermore, OpenIntersection dynamically manages this communication spectrum based on application priority, network load, and policy configurations encoded at the edge. This hybrid model achieves a balance between efficiency and responsiveness, aligning bandwidth consumption and computational effort with the urgency and contextual importance of the data being exchanged.

Moreover, OpenIntersection serves as a foundation for more advanced applications in connected driving ecosystems. It can support predictive analytics by integrating historical traffic patterns with real-time data, enabling anticipatory control strategies. It also lays the groundwork for cooperative perception, where vehicles and infrastructure jointly construct a shared, dynamic representation of the intersection environment.

Incorporating insights from the broader field of edge computing for intelligent transportation, OpenIntersection reflects the trend toward decentralized intelligence, as described in prior foundational work on vehicular edge computing. Specifically,

OpenIntersection aligns with the principles outlined by Shi et al. [10], who argue that latency-sensitive applications, such as those in vehicular safety, cannot depend solely on cloud services due to delay variability and bandwidth constraints. Open-Intersection implements this vision by leveraging edge devices at the intersection to process, aggregate, and act on data locally, thereby ensuring responsiveness and scalability. Furthermore, by offloading computation to the network edge, OpenIntersection reduces the burden on vehicle-side resources and improves energy efficiency—two objectives emphasized in vehicular edge computing architectures.

4.3 The Basics of V2X Protocols

V2X communication is classified into several categories, each serving a unique function in enhancing safety, efficiency, and coordination on the road.

Vehicle-to-Vehicle (V2V) This mode enables direct communication between nearby vehicles. Consider a multilane freeway scenario during a sudden snow squall. Vehicle A, driving ahead, detects a patch of black ice using onboard sensors and stability control systems. It transmits a warning message containing GPS location, severity of wheel slip, and a timestamp to following vehicles. Vehicle B, several hundred meters behind and currently unable to detect the hazard visually or physically, receives the alert and preemptively reduces speed, enabling the driver assistance system to prepare for reduced traction. This kind of early warning system allows chain-reaction collisions to be avoided, especially in poor visibility conditions.

Vehicle-to-Infrastructure (V2I) This communication occurs between vehicles and road infrastructure such as traffic signals and variable message signs. Take, for instance, a coordinated traffic management system in a smart city. Vehicle C is approaching an urban intersection where the traffic signal is equipped with a roadside unit broadcasting SPaT messages. Based on these messages, the vehicle's control system computes that the green light will end in 3.2 seconds. Instead of accelerating to "beat the light," the vehicle smoothly decelerates, conserving energy and improving traffic flow. Meanwhile, the traffic light system detects a high-priority bus running behind schedule and extends the green phase by 5 seconds to support public transit reliability. This dynamic coordination improves safety and system-wide efficiency.

Vehicle-to-Pedestrian (V2P) This mode allows vehicles to communicate with pedestrians carrying connected devices. Imagine a school zone equipped with beacon systems and wearable V2X tags for children. A child wearing a tag begins to cross the road between parked cars—out of sight from oncoming traffic. The tag broadcasts the child's location and movement vector to nearby vehicles. Vehicle D, a partially autonomous car approaching the zone, instantly receives the alert and overrides its acceleration plan to initiate a full stop. Simultaneously, a visual alert is issued inside the cabin to notify the human driver. This scenario illustrates how V2P can offer a critical layer of protection for vulnerable road users in unpredictable environments.

Vehicle-to-Network (V2N) This mode connects the vehicle to broader network services via cellular infrastructure. Consider a regional wildfire emergency in a mountainous area. Evacuation routes are changing by the minute due to shifting fire fronts and road closures. Vehicle E queries a cloud-based emergency operations center via 5G and receives a real-time evacuation map tailored to its location. The vehicle's navigation system integrates the data and continuously updates its path, while broadcasting its estimated route and timing to other vehicles and traffic control systems via V2V and V2I. This allows dynamic coordination of evacuation traffic and reduces bottlenecks along narrow mountain passes.

Examples Cooperative perception systems enhance the spatial awareness of autonomous vehicles by leveraging shared sensor information through V2X channels. Ren et al. introduce an interruption-aware cooperative perception framework that maintains perception performance in adverse communication conditions, proposing a robust architecture to cope with dynamic bandwidth and latency [14]. Han et al. provide a broad survey on collaborative perception, covering techniques, datasets, and practical challenges in multi-agent perception scenarios [4]. Huang et al. further analyze recent advances and limitations in V2X cooperative perception, offering insights into fusion strategies, edge computing support, and communication constraints [8]. Zhang et al. present EMP, an edge-assisted multi-vehicle perception framework that utilizes roadside edge devices to aggregate and process visual data, enabling richer environmental awareness [20]. Collision avoidance is another primary safety application of V2X communications. Deng et al. propose a cooperative overtaking scheme using C-V2X for safe and efficient collision avoidance during lane changes, supported by game-theoretic modeling and real-time communication strategies [3]. Wu et al. present "SafeCross," a cooperative decision framework that uses infrastructure support to determine safe crossing behaviors at intersections, accounting for occlusions and dynamic agents [16]. Innovative architectures and frameworks are also emerging to enhance communication reliability and system scalability. Wu et al. propose "Tentacles," a middleware designed to support multi-network communication for infrastructure-assisted cooperative driving, addressing redundancy, latency, and load-balancing in vehicular networks [17]. He and Shi envision transformative intersection designs that leverage connectivity and edge computing to orchestrate multi-agent cooperation in urban environments [5].

Artificial intelligence and large multimodal models are increasingly being applied to cooperative autonomous driving. You et al. propose V2X-VLM, an end-to-end system that leverages large vision-language models for collaborative driving tasks, demonstrating the feasibility of combining language-driven reasoning with perception-sharing across vehicles [18] (Fig. 4.1).

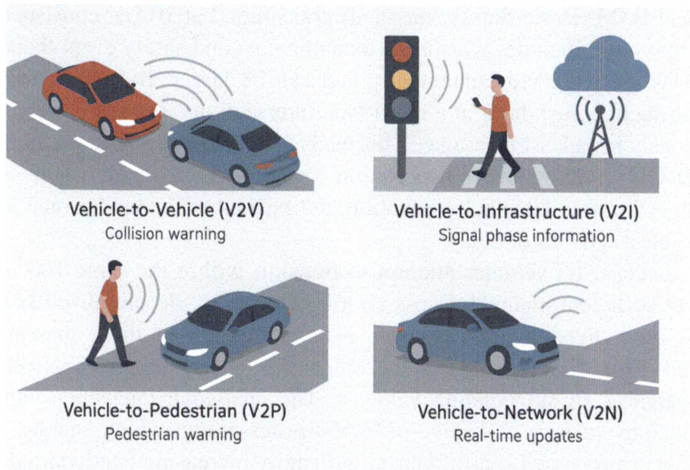

Fig. 4.1 An illustration showing four types of V2X communication

4.4 Challenges in V2X Communication

Despite its promise, the deployment and operation of Vehicle-to-Everything (V2X) communication systems face numerous technical and systemic challenges. These limitations must be addressed to ensure V2X systems can support the stringent reliability, safety, and latency requirements of autonomous driving and intelligent transportation applications.

Theoretical and empirical studies also point to the challenges of managing access control, data integrity, and teleoperation. Zhang et al. design AC4AV, a dynamic access control system tailored to connected and autonomous vehicle contexts, ensuring data confidentiality and selective sharing [19]. Lu and Shi discuss the role of teleoperation as a safety fallback and control extension for connected vehicles, detailing technical frameworks for operator-vehicle coordination over wireless networks [11].

Broader survey and review articles synthesize the overall direction of connected autonomous driving. Montanaro et al. analyze various connected driving use cases, such as platooning, intersection negotiation, and cooperative lane changes, contextualizing them within regulatory and technological readiness levels [13]. Wang et al. provide a comprehensive overview of networking and communication strategies in autonomous driving, focusing on communication models, protocol stacks, and open research questions [15]. He et al. outline the collaborative capabilities of C-V2X and evaluate its real-world implementation and performance in diverse scenarios [6]. Luo et al. address content distribution in vehicular edge computing via their EdgeVCD framework, which optimizes data dissemination using intelligent algorithms [12].

A typical BSM (Basic Safety Message), transmitted at 10 Hz, consists of about 200–300 bytes and includes real-time kinematic data and safety event flags. CAMs and DENMs are triggered contextually and can be larger. In high-density traffic scenarios—such as rush hour at a multi-lane intersection—dozens of vehicles may simultaneously broadcast messages. This leads to spectrum congestion, particularly in the 5.9 GHz Intelligent Transportation Systems (ITS) band, where available bandwidth is limited. Channel contention and packet collisions become frequent, increasing message loss or delay.

For instance, if 60 vehicles attempt to transmit within the same 100 ms interval without sufficient channel access coordination (e.g., decentralized congestion control or time division multiplexing), a large fraction of those messages may be dropped. This directly impacts the efficacy of safety applications relying on timely awareness of surrounding vehicles. The challenge intensifies with future V2X extensions such as collective perception messaging, which enables vehicles to share raw or processed sensor data, significantly increasing bandwidth demands. Without more efficient spectrum usage strategies, dynamic prioritization protocols, and scalable media access control schemes, V2X systems may fail to deliver timely data precisely when it is most critical. Latency: Current safety-related V2X applications benefit from low-latency communication, but they are not strictly time-critical in the sense of requiring end-to-end latencies below 10 milliseconds. Most deployed use cases—such as forward collision warnings and emergency electronic brake light alerts—are designed to enhance driver and vehicle situational awareness by providing additional context from outside the line-of-sight, rather than to directly control vehicle actuation in real time. These applications typically tolerate latencies in the range of 100–300 milliseconds. However, latency remains a critical performance metric because reductions in message delay can substantially increase the effective time window for a driver or automated system to respond.

In another scenario, if Vehicle A detects an obstacle and transmits a warning message via V2V, the timeliness of that message's delivery to Vehicle B affects whether B can integrate the information into its own planning process. A delay of 200 milliseconds might be acceptable at low urban speeds, but at highway velocities, it could significantly reduce the margin for evasive action. Additionally, latency becomes more problematic in congested or interference-prone environments, such as urban canyons, where signal multipath and channel contention are common. Future use cases such as cooperative collision avoidance or coordinated intersection negotiation may require lower and more deterministic latency. While today's V2X systems do not yet demand sub-10 ms performance, developing the network capability to support such requirements—through techniques like 5G URLLC and decentralized edge computing—will be essential for more tightly coupled cooperative behaviors in autonomous systems.

Bandwidth Limitations Each connected vehicle generates multiple classes of data intended for cooperative driving, situational awareness, and, in some cases, infotainment. The most commonly transmitted messages include low-level state information such as a vehicle's position, velocity, acceleration, heading, and size.

These messages may also include intent indicators, such as whether the turn signal is activated, the brake is applied, or a lane change is in progress. More advanced V2X systems can support the transmission of raw sensor data—for instance, LiDAR point clouds or radar detections—or processed perception data, such as identified objects and their motion predictions. Additionally, decision-level information such as planned trajectories, stopping intentions, or yield behavior may be shared between vehicles and infrastructure to facilitate cooperative maneuvers, such as merging at high speeds or negotiating complex intersections. The breadth of shared data depends on system architecture, communication bandwidth, and application requirements.

To mitigate these risks, modern V2X systems rely on a security credential management system (SCMS), which uses Public Key Infrastructure (PKI) to authenticate messages. Each RSU and OBU contains a hardware security module (HSM) that stores cryptographic keys and performs digital signature operations. Messages are signed using a short-term certificate (pseudonym) to ensure integrity and authenticity while protecting user privacy. However, implementing PKI at scale raises significant challenges. Certificate authorities must be globally trusted and synchronized, and certificate revocation mechanisms must function in a highly dynamic and mobile environment. Moreover, the overhead of cryptographic signing and verification can introduce latency and computational load, particularly on embedded hardware.

In practice, secure V2X communication must strike a balance between message authentication, privacy, scalability, and real-time responsiveness. Designing systems that prevent message tampering while supporting low-latency communication remains one of the most complex aspects of deploying V2X technologies.

Standardization and Interoperability For V2X systems to function at scale, seamless communication between heterogeneous vehicles and infrastructure—regardless of vendor, geographic region, or communication stack—is essential. However, the current V2X ecosystem remains fragmented across multiple dimensions. At the physical and MAC layers, IEEE 802.11p (DSRC) and 3GPP C-V2X use incompatible radio technologies. Even when harmonized message formats such as SAE J2735 or ETSI ITS-G5 are adopted, inconsistencies in implementation, certification, and deployment policies frequently cause interoperability failures.

Furthermore, most V2X deployments today are designed around fixed-function message types and rigid communication primitives, such as broadcasting Basic Safety Messages (BSMs) at fixed intervals. These static designs lack the flexibility needed to accommodate emerging use cases—such as dynamic intersection management or multi-agent coordination—which require vehicles and infrastructure to exchange semantic-level, context-specific information.

The OpenIntersection project addresses this gap by proposing a software-defined interoperability framework in which roadside units expose programmable interfaces, allowing vehicles to negotiate intersection traversal using declarative policies and abstracted message semantics. Instead of relying solely on pre-defined message formats, vehicles can engage in higher-level protocols mediated by the intersection controller. This model also accommodates diverse V2X stacks by translating between protocol-specific encodings and shared abstract representations.

Such software-defined approaches offer a path toward semantic interoperability without requiring full standardization at every layer of the stack. However, their adoption will depend on widespread agreement on common APIs, execution sandboxes, and trust models. Until such frameworks are standardized and integrated with existing infrastructure, the V2X ecosystem will continue to suffer from fragmented deployment and limited cross-vendor compatibility.

These challenges—low latency requirements, limited bandwidth, cybersecurity threats, and lack of global standardization—present critical barriers to the scalability and reliability of V2X systems. Overcoming them will require coordinated innovation across communication technologies, policy frameworks, and vehicle design, backed by strong cross-sector collaboration between automakers, governments, and technology providers.

4.5 Push and Pull Services in V2X Communication

Push and pull are two foundational service models that define how information flows in a networked system. In the context of V2X communication, these models do not refer to any specific V2X standard or protocol, but rather describe general paradigms of data delivery that can be supported by V2X infrastructure and applications. Their significance lies in determining not only what data is transmitted and when, but also how resources such as bandwidth, computation, and attention are allocated in increasingly complex vehicle ecosystems.

Push services are characterized by proactive dissemination of data. The sender determines the content, timing, and destination scope of the message without any solicitation from recipients. These messages are typically broadcast or multicast to all nearby vehicles or infrastructure units within a communication radius. Push services are advantageous in cases where the data is likely to be useful to many recipients simultaneously, or where the urgency of the message makes waiting for a request impractical. Examples include general-purpose announcements (e.g., traffic congestion alerts), safety warnings (e.g., sudden braking or obstacle detection), or infrastructure-generated data (e.g., signal phase and timing). In push models, the responsibility of deciding message relevance is transferred to the receivers, which must filter incoming data in real time.

Pull services invert this model: communication is initiated by the receiver, which explicitly requests data from a source. This is more resource-efficient in situations where data utility is conditional on a vehicle's specific state, location, or intention. Pull services are well-suited for differentiated information services, where only a subset of available data is required. For example, a vehicle approaching an unfamiliar area may query for high-resolution local maps, or request the control policy of an upcoming intersection. In cloud-assisted navigation, a vehicle may request optimal routing information given current congestion and historical flow trends. Pull requests can be unicast, allowing precise delivery of information and preserving channel bandwidth.

Push and pull models also differ in their temporal logic. Push is appropriate for periodically updated or event-triggered data that loses value quickly if delayed, while pull is suitable for static or semi-static data that remains valid until conditions change. An important architectural distinction is that push-based systems typically require robust congestion management and message prioritization, while pull-based systems require responsive, addressable infrastructure capable of fulfilling heterogeneous requests with minimal latency.

In the context of future V2X systems—especially those incorporating programmable infrastructure like OpenIntersection—the interplay between push and pull becomes even more critical. Infrastructure may support multiple tiers of service, where low-level status messages are pushed continuously, while higher-level negotiation parameters (e.g., intersection traversal slots, cooperative intent synchronization) are exchanged via pull requests. This separation supports modular service composition, improves scalability, and aligns data delivery with application-specific timing and relevance constraints.

Effective V2X system design will require policies for deciding when a message should be pushed versus when it should be pulled, possibly based on dynamic vehicle state, current network load, or historical service outcomes. As vehicular networks evolve toward greater autonomy and context sensitivity, understanding and designing for the distinction between push and pull models will be essential for robust and intelligent communication systems.

4.6 V2X Applications

Figure 4.2 depicts a scene where two vehicles are approaching a four-way intersection with no traffic signals on a foggy morning. Each car is equipped with a V2V system that continuously broadcasts its position, speed, and heading. As they near the intersection from perpendicular directions, their onboard systems calculate a potential collision risk. A warning is issued to both drivers—or to the automated driving systems—prompting immediate braking. Thanks to the early communication, a crash is avoided even though neither vehicle had direct line-of-sight to the other due to the weather conditions.

Figure 4.3 depicts a scene where a connected vehicle is approaching a traffic light in a busy downtown area,. The traffic signal is part of a smart infrastructure system that transmits its current phase and timing (SPaT) data. The vehicle receives a message indicating that the green light will end in just over three seconds. Instead of accelerating, the vehicle gently slows down, improving fuel efficiency and safety. At the same time, the intersection control system detects that a public transit bus is behind schedule and extends the green phase by five seconds to help it stay on route.

Figure 4.4 depicts a scene where children wearing V2X-enabled smart devices begin crossing the street between parked cars during afternoon dismissal at a local

Fig. 4.2 Vehicle to Vehicle application

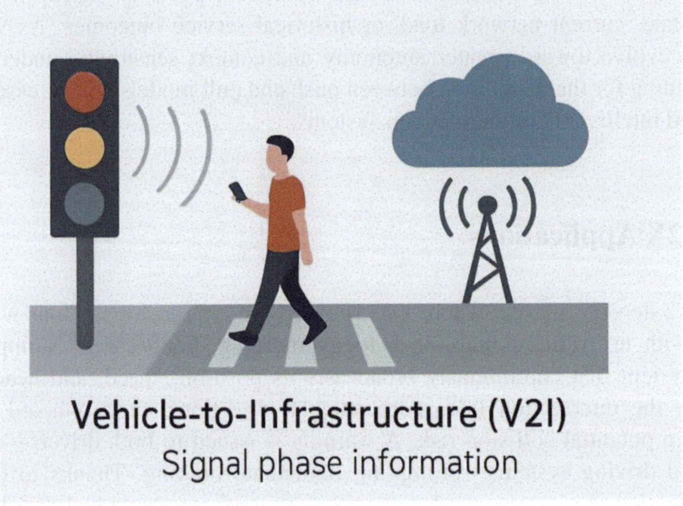

Fig. 4.3 Vehicle to Infrastructure application

elementary school. A nearby autonomous vehicle, which cannot visually detect the children due to the obstruction, receives a signal from one of the wristbands indicating a child's precise location and movement direction. Instantly, the vehicle slows to a stop and flashes a warning on the dashboard for the supervising adult in the front seat. The child safely crosses the street without ever being in the vehicle's line of sight.

Fig. 4.4 Vehicle to Pedestrian application

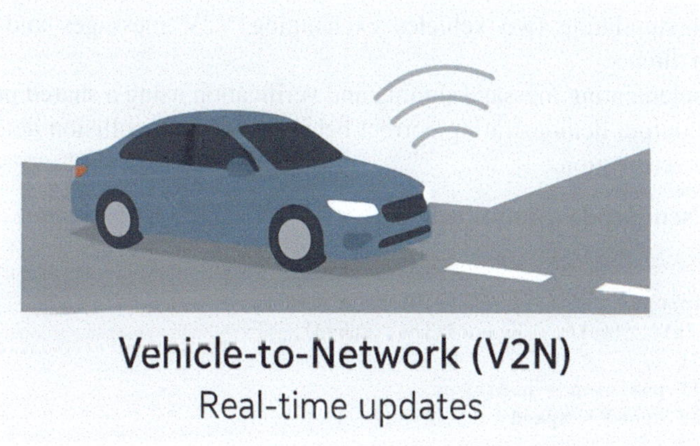

Fig. 4.5 Vehicle to network application

Figure 4.5 depicts a scene where a vehicle with V2N capability receives an alert from a regional emergency operations center via 5G about a fast-moving wildfire in the vicinity. The alert includes an updated evacuation route based on real-time road closures and fire progression data. The vehicle's navigation system adapts immediately, rerouting it away from danger. Simultaneously, the vehicle broadcasts its new route and estimated arrival time to nearby traffic control systems, allowing for coordinated flow management along narrow escape roads.

4.7 Hands-on Exercises

4.7.1 Implement a Basic V2V Communication System with Collision Detection and Message Authentication

Task Develop a basic Vehicle-to-Vehicle (V2V) communication system using a publish-subscribe architecture. Each vehicle broadcasts its state (position, speed, heading), and nearby vehicles use this information to detect potential collision risks. Extend the system with message signing to ensure data integrity.

Object Simulate two vehicles approaching an uncontrolled intersection. Implement logic to (1) broadcast and subscribe to state data, (2) compute collision warnings based on relative position and speed, and (3) authenticate messages using cryptographic hashing.

Deliverable Submit your implementation in Python or C++, including:

1. A script simulating two vehicles exchanging V2V messages and detecting collision threats.
2. Code implementing message signing and verification using a shared private key.
3. Console output demonstrating correct behavior for both collision detection and message verification.

Example Pseudocode (Simplified)

```
class Vehicle:
    def __init__(self, id, position, speed):
        self.id = id
        self.position = position
        self.speed = speed

    def broadcast(self):
        return {'id': self.id, 'position': self.position, 'speed': self.speed
            }

def detect_collision(v1, v2):
    if abs(v1['position'] - v2['position']) < 5 and abs(v1['speed'] - v2['
        speed']) < 5:
        return "Collision␣Warning!"
    return "Safe"

# Message signing
import hashlib
def sign(msg, key): return hashlib.sha256((msg + key).encode()).hexdigest()
def verify(msg, sig, key): return sign(msg, key) == sig
```

Note: Validate your system within a simple simulated loop or discrete-time framework. Optionally extend to multi-vehicle interactions or integrate with a ROS-based simulator.

Recommended Papers to Read

1. He, Y., & Shi, W. (2024). A Vision for Transformative Intersections. Computer, 57(12), 58–68. [5]
2. Wu, T., Wang, S., Bao, Y., & Shi, W. (2024, October). Tentacles: A Middleware with Multi-Network Communication Reliability for Vehicle-Infrastructure Cooperative Autonomous Driving. In 2024 IEEE 100th Vehicular Technology Conference (VTC2024-Fall) (pp. 1–8). IEEE. [17]
3. He, Y., Wu, B., Dong, Z., Wan, J., & Shi, W. (2023). Towards c-v2x enabled collaborative autonomous driving. IEEE Transactions on Vehicular Technology, 72(12), 15450–15462. [6]

References

1. Bagheri, H., et al. (2021). 5G NR-V2X: Toward connected and cooperative autonomous driving. *IEEE Communications Standards Magazine*, *5*(1), 48–54.
2. Cinque, E., et al. (2020). V2X communication technologies and service requirements for connected and autonomous driving. In *2020 AEIT International Conference of Electrical and Electronic Technologies for Automotive (AEIT AUTOMOTIVE)* (pp. 1–6). IEEE.
3. Deng, R., Di, B., & Song, L. (2019). Cooperative collision avoidance for overtaking maneuvers in cellular V2X-based autonomous driving. *IEEE Transactions on Vehicular Technology*, *68*(5), 4434–4446.
4. Han, Y., et al. (2023). Collaborative perception in autonomous driving: Methods, datasets, and challenges. *IEEE Intelligent Transportation Systems Magazine*, *15*(6), 131–151.
5. He, Y., & Shi, W. (2024). A vision for transformative intersections. *Computer*, *57*(12), 58–68.
6. He, Y., et al. (2023). Towards C-V2X enabled collaborative autonomous driving. *IEEE Transactions on Vehicular Technology*, *72*(12), 15450–15462.
7. Hobert, L., et al. (2015). Enhancements of V2X communication in support of cooperative autonomous driving. *IEEE Communications Magazine*, *53*(12), 64–70.
8. Huang, T., et al. (2023). *V2X cooperative perception for autonomous driving: Recent advances and challenges*. Preprint. arXiv:2310.03525.
9. Kutila, M., et al. C-V2X supported automated driving. In *2019 IEEE International Conference on Communications Workshops (ICC Workshops)* (pp. 1–5). IEEE.
10. Liu, S., et al. (2019). Edge computing for autonomous driving: Opportunities and challenges. *Proceedings of the IEEE*, *107*(8), 1697–1716. ID: 1. https://doi.org/10.1109/JPROC.2019.2915983
11. Lu, S., Zhong, R., & Shi, W. (2022). Teleoperation technologies for enhancing connected and autonomous vehicles. In *2022 IEEE 19th International Conference on Mobile Ad Hoc and Smart Systems (MASS)* (pp. 435–443). IEEE.
12. Luo, Q., et al. (2020). EdgeVCD: Intelligent algorithm-inspired content distribution in vehicular edge computing network. *IEEE Internet of things Journal*, *7*(6), 5562–5579.
13. Montanaro, U., et al. (2019). Towards connected autonomous driving: review of use-cases. *Vehicle System Dynamics*, *57*(6), 779–814.
14. Ren, S., et al. (2024). Interruption-aware cooperative perception for V2X communication-aided autonomous driving. *IEEE Transactions on Intelligent Vehicles*.

15. Wang, J., Liu, J., & Kato, N. (2018). Networking and communications in autonomous driving: A survey. *IEEE Communications Surveys and Tutorials*, *21*(2), 1243–1274.
16. Wu, B., et al. (2022). To turn or not to turn, safecross is the answer. In *2022 IEEE 42nd International Conference on Distributed Computing Systems (ICDCS)* (pp. 414–424). IEEE.
17. Wu, T., et al. (2024). Tentacles: A middleware with multi-network communication reliability for vehicle-infrastructure cooperative autonomous driving. In *2024 IEEE 100th Vehicular Technology Conference (VTC2024-Fall)* (pp. 1–8). IEEE.
18. You, J., et al. (2024). *V2X-VLM: End-to-end V2X cooperative autonomous driving through large vision-language models*. Preprint. arXiv:2408.09251.
19. Zhang, Q., et al. (2020). AC4AV: A flexible and dynamic access control framework for connected and autonomous vehicles. *IEEE Internet of Things Journal*, *8*(3), 1946–1958.
20. Zhang, X., et al. (2021). EMP: Edge-assisted multi-vehicle perception. In *Proceedings of the 27th Annual International Conference on Mobile Computing and Networking* (pp. 545–558).

Chapter 5
Perception Algorithms

5.1 Introduction to Perception in Autonomous Vehicles

Perception in autonomous vehicles (AVs) involves the intricate process of interpreting environmental data to facilitate safe, efficient, and reliable navigation in complex driving scenarios. Serving as one of the fundamental pillars supporting autonomous driving systems, perception enables vehicles to perform several critical functions. This includes detecting a wide variety of static and dynamic objects, accurately classifying and tracking these objects over time, and interpreting their movements to forecast future states. Beyond object recognition, perception encompasses the identification and interpretation of lane markings to ensure proper vehicle positioning, recognition and understanding of traffic signs to ensure compliance with road regulations, and the prediction of complex environmental dynamics such as pedestrian and vehicle behaviors.

To accomplish these sophisticated tasks, modern AV perception systems leverage an integrated network of sensors, each offering distinct advantages and contributing unique insights. Cameras capture high-resolution visual data ideal for detailed object identification and semantic understanding. LiDAR sensors produce precise 3D point clouds, allowing for accurate depth perception and reliable detection of objects at varying distances. Radar sensors excel at detecting object speed and distance under challenging weather conditions, while ultrasonic sensors are primarily utilized for short-range detection tasks, such as parking and close-proximity maneuvers.

The integration and synthesis of data from these diverse sensors involve advanced data fusion techniques and machine learning algorithms, including deep neural networks and probabilistic models, to generate real-time, robust, and comprehensive models of the vehicle's surroundings. This multifaceted approach ensures that autonomous vehicles can respond appropriately to dynamic and unpredictable environments, enhancing safety and operational effectiveness.

W. Shi, Y. He, *Introduction to Autonomous Driving*,
https://doi.org/10.1007/978-3-031-99485-2_5

The critical nature of perception requires continuous, accurate analysis of incoming sensor data. Raw sensor signals must be meticulously processed to extract meaningful insights. Core perception tasks encompass object detection and tracking, semantic segmentation, lane detection and tracking, traffic sign recognition and tracking, anomaly detection, obstacle avoidance, and precise environmental mapping. Achieving robust and reliable perception is crucial, as it allows autonomous vehicles to make safe and informed decisions, particularly in complex and dynamic real-world driving conditions.

5.2 Computer Vision Algorithms for Perception

Computer vision algorithms significantly enhance an autonomous vehicle's capability to interpret and make sense of complex visual environments, closely emulating human vision but extending far beyond it in terms of speed, precision, and consistency. These algorithms form the cornerstone of vehicle perception, enabling the identification, classification, tracking, and prediction of various entities in the vehicle's surroundings.

They encompass a wide spectrum of approaches, beginning with classical methods such as edge detection, image thresholding, and feature extraction algorithms like the Hough Transform and Histogram of Oriented Gradients (HOG). These intuitive techniques are often effective for simpler tasks like basic lane detection or the identification of well-defined shapes and patterns.

However, as driving environments become increasingly complex and unpredictable, more sophisticated methods are required. Modern deep learning frameworks, including convolutional neural networks (CNNs), recurrent neural networks (RNNs), and transformer-based architectures, have dramatically improved the capabilities of computer vision systems. CNNs excel at capturing hierarchical spatial features essential for detailed object detection, lane recognition, and semantic segmentation. RNNs and transformers, by contrast, are particularly effective in scenarios involving sequential data or temporal dynamics, such as tracking objects over consecutive frames or predicting pedestrian and vehicle behaviors.

5.2.1 *Object Detection and Tracking*

You Only Look Once (YOLO) [36] is a widely-used real-time object detection algorithm initially proposed by Joseph Redmon and colleagues in 2016. YOLO gained traction due to its innovative approach that framed object detection as a single regression problem rather than a combination of classification and localization subtasks. Unlike traditional algorithms such as R-CNN and Fast R-CNN, which rely on region proposals and multiple passes over the image, YOLO partitions the image into a grid and simultaneously predicts bounding boxes and class probabilities for each cell. This direct, end-to-end methodology enables real-time performance even on modest hardware, making it well-suited for autonomous systems and robotics.

The architecture of YOLO employs convolutional layers for feature extraction followed by fully connected layers that output bounding box coordinates and class scores. Each grid cell predicts multiple bounding boxes, each defined by center coordinates, dimensions, and confidence scores, along with conditional class probabilities. YOLO's speed advantage stems from this unified model structure, though it may struggle with detecting small or densely packed objects due to its grid-based representation.

YOLO's capabilities are further extended in subsequent iterations, including YOLOX [12], which introduced an anchor-free detection head and decoupled classification and regression branches, enhancing both training stability and accuracy.

Faster R-CNN [37] is a seminal object detection framework that significantly improved both speed and accuracy over its predecessors by integrating region proposal and classification into a single, end-to-end trainable system. The model introduces the Region Proposal Network (RPN), a fully convolutional network that shares features with the detection network to generate region proposals efficiently. This removes the need for computationally expensive external proposal methods like Selective Search, which were used in earlier models such as Fast R-CNN. By coupling the RPN with Fast R-CNN's region-based classifier, Faster R-CNN provides a unified pipeline where region proposal generation and object detection occur simultaneously. The major problem it solves is the bottleneck in detection speed and inefficiencies associated with multi-stage detection pipelines. Its contribution lies in enabling near real-time detection without sacrificing accuracy, setting a new standard for high-precision object detectors used in various computer vision tasks, including autonomous driving and surveillance.

To enable persistent identification of moving objects across video frames, object detection must be paired with tracking algorithms. The Simple Online and Real-time Tracking (SORT) algorithm exemplifies this integration. SORT uses a Kalman filter to predict future states of detected objects and the Hungarian algorithm for data association between frame-to-frame detections. The system initializes state vectors with parameters such as position and velocity and updates them based on new observations. Tracks are created for unmatched detections and removed after a period of inactivity [49].

Further extensions like Deep SORT incorporate appearance embeddings to improve track consistency through occlusions and interactions.

More recently, transformer-based approaches have merged detection and tracking into unified architectures. TrackFormer [26] introduces object queries to represent tracks, allowing end-to-end learning of tracking without explicit data association. Similarly, TransTrack [43] leverages attention mechanisms to maintain track continuity over frames by associating detections through spatiotemporal embeddings.

Other approaches such as decision-based tracking formulate tracking as a sequential decision-making process [52], optimizing long-term associations rather than frame-local matching.

Benchmarking efforts have shaped the evaluation of tracking algorithms. Wu et al. [51] provide comprehensive assessments on online tracking datasets and emphasize a crucial limitation in early evaluation protocols: tracking performance was often overestimated due to reliance on ground-truth or near-perfect detections. In practice, object detection is noisy, with frequent missed detections or false positives. Wu et al. highlighted how trackers that performed well in ideal conditions frequently degraded under real-world scenarios, where detector imperfections introduced significant errors in tracking association. Their benchmark introduced realistic detection outputs to better isolate and evaluate tracker robustness in practical applications.

Comaniciu et al. [8] introduced kernel-based tracking, a class of algorithms that represent the tracked object by a probability density function, typically modeled using color histograms or other appearance features. The most prominent technique they proposed was based on the mean-shift algorithm, which iteratively moves the search window toward the maximum similarity between the current frame and the target model. This method is computationally efficient and particularly robust against partial occlusions and changes in object appearance. It addressed the challenge of real-time object localization without relying on exhaustive search or complex motion models. The contribution of their work lies in formalizing kernel-based tracking under a non-parametric framework and demonstrating its effectiveness across a variety of video sequences, which subsequently influenced many real-time and embedded tracking systems.

Survey works by Luo et al. and Yilmaz et al. [24, 57] offer structured and in-depth examinations of the evolution of multi-object tracking. Luo et al. provide a taxonomy categorizing tracking algorithms into online and offline methods, and further divide techniques based on the reliance on detection, motion models, and learning-based components. They also discuss metrics such as MOTA and IDF1 and analyze their implications for benchmarking. Yilmaz et al. give a broader historical perspective, outlining the challenges in tracking under occlusions, illumination changes, and clutter, and compare generative and discriminative tracking paradigms. Together, these surveys not only contextualize the technical progress in the field but also highlight enduring challenges and open research questions.

Traffic sign detection and tracking is a fundamental task in the perception stack of autonomous vehicles (AVs), as it directly influences decision-making related to speed regulation, lane merging, and priority at intersections. The reliable identification and interpretation of traffic signs ensure that the vehicle adheres to road rules and reacts appropriately to dynamic driving environments. An effective perception system must not only recognize a wide variety of traffic signs across different regions and jurisdictions but also cope with partial occlusions, faded signs, and temporary traffic signals, such as construction warnings. Real-time operation and minimal latency are essential, as delayed or missed detections can compromise safety and performance.

Benchmarking datasets and traditional systems laid the groundwork for modern methods. The German Traffic Sign Detection Benchmark (GTSDB), introduced by Houben et al., provided one of the earliest comprehensive resources for evaluating traffic sign detection models in real-world scenarios [16]. The dataset includes challenging samples with occlusions, varying weather conditions, and different viewing angles, enabling researchers to assess the generalization capacity of their models. Its significance lies not only in dataset size and diversity but also in standardizing evaluation metrics, which facilitated direct comparisons among competing approaches and accelerated progress in the field.

Early detection systems such as the one proposed by Yang et al. [56] implemented real-time traffic sign recognition on embedded hardware using color and shape-based heuristics alongside support vector machines (SVMs). This work was pivotal in demonstrating that traffic sign detection could be feasible on resource-constrained platforms typical of early AV prototypes. The system addressed issues of low computational power and the need for robust classification under varying illumination conditions, laying the conceptual foundation for real-time embedded deployment in AVs.

Building on these foundations, contemporary models have leveraged deep learning to dramatically improve detection accuracy and adaptability. Wang et al. [45] enhanced the YOLOv5 architecture by incorporating multi-scale feature fusion, attention modules, and optimized loss functions to better capture small and distant traffic signs. Their model significantly improved precision and recall metrics, particularly for challenging edge cases like partially obscured or low-resolution signs. This contribution is crucial for AVs navigating complex environments where missed detections can result in traffic violations or unsafe maneuvers.

In a related study, Soylu and Soylu [41] performed a systematic comparison of YOLOv8 variants on the Robotaxi Full-Scale AV Challenge dataset. Their evaluation emphasized the trade-offs between model accuracy, inference speed, and computational resource demands. By quantifying the performance of different YOLOv8 models under competition-like constraints, their study provides actionable insights for AV developers selecting models based on specific deployment requirements, such as latency bounds or hardware limitations.

Transformer-based architectures have also been adapted for this domain to enhance context understanding and spatial reasoning. Zhang et al. [61] introduced TSD-DETR, a lightweight variant of the DEtection TRansformer tailored to traffic sign detection. Unlike general-purpose DETR models, TSD-DETR employs scale-aware attention and a compact backbone network to ensure high detection accuracy without exceeding real-time computational budgets. Its ability to detect small and rare traffic signs in complex visual scenes makes it highly applicable to urban driving environments where regulatory signage is critical yet often subtle.

5.2.2 Detection Techniques

Traditional computer vision approaches have historically served as the foundational framework for detecting traffic signs in visual data captured by vehicle-mounted cameras. These approaches rely on a structured sequence of image processing operations aimed at identifying visually distinctive features that correspond to traffic signs. The process typically begins with color space transformations; for instance, converting the original RGB image into HSV (Hue, Saturation, Value) format allows for more effective separation of color features, making it easier to detect signs that use standard colors like red, blue, and yellow. This step is particularly useful because the HSV color space separates chromatic content (hue) from intensity (value), making the system more robust to shadows and lighting variations.

Following color filtering, edge detection techniques such as the Canny edge detector are applied to delineate object boundaries and highlight the sharp transitions in intensity typically found at the edges of traffic signs. These edges often correspond to geometric shapes like rectangles, circles, and triangles, which are characteristic of various traffic signs. The next stage involves contour detection and shape analysis, where the system identifies closed contours and matches them against predefined templates representing known sign geometries. These candidate regions are then subjected to further analysis to verify whether they correspond to traffic signs.

Classification is subsequently performed using handcrafted feature descriptors such as Histograms of Oriented Gradients (HOG), which capture the distribution of edge orientations within a region. These features are compared against a reference database of known traffic sign patterns using methods like support vector machines (SVMs) or template matching algorithms. Although these traditional pipelines are relatively lightweight and computationally efficient, they suffer from limitations in terms of generalizability and robustness. Their performance tends to degrade significantly under conditions involving variable lighting, complex or cluttered backgrounds, partial occlusions, and low-resolution imagery, all of which are common in real-world driving scenarios. Despite their limitations, these methods provided the initial conceptual structure for later learning-based approaches, helping define baseline detection strategies.

With the advent of deep learning, convolutional neural networks (CNNs) have significantly advanced the field. CNN-based detectors automate the feature extraction process, allowing models to learn hierarchical patterns directly from labeled data. State-of-the-art object detectors like YOLO (You Only Look Once) and SSD (Single Shot MultiBox Detector) offer robust, end-to-end frameworks for both classification and localization. These models are trained on large-scale annotated datasets like the German Traffic Sign Recognition Benchmark (GTSRB) and the LISA Traffic Sign Dataset, which include diverse conditions such as occlusions, motion blur, and low illumination. Unlike traditional pipelines, CNN-based systems demonstrate better generalization to new scenes and unseen sign types. Moreover, their ability to process entire images in a single forward pass supports real-time inference essential for embedded AV systems operating under strict latency constraints.

Keiron O'Shea and Ryan Nash [30] provided a foundational introduction to convolutional neural networks, elucidating their architecture and training mechanisms. Their work demystified how convolutional layers function to extract spatial hierarchies of features, how pooling layers contribute to translational invariance, and how fully connected layers integrate global information for classification. By formalizing the core structure and intuition of CNNs, this paper enabled broader adoption and adaptation of deep learning models for vision tasks, including autonomous driving perception modules. It serves as a cornerstone for educational and engineering teams aiming to implement or optimize neural network architectures.

Zou et al. [64] introduced the Segment Anything Model (SAM), a significant leap in general-purpose segmentation tools. SAM unifies semantic, instance, and panoptic segmentation under a single framework using a prompt-based mechanism, enabling segmentation of any object class without requiring class-specific retraining. Trained on over one billion masks, SAM demonstrated strong zero-shot generalization across domains, including urban and vehicular environments. Its significance lies in its potential to serve as a universal segmentation backbone for AV systems, offering rapid adaptability and reduced training cost in diverse deployment contexts.

Shi et al. [39] proposed LLMFormer, which bridges vision and language models to support open-vocabulary semantic segmentation. By aligning image features with textual embeddings via transformer-based architectures, LLMFormer enables the recognition of novel or rare object classes based on natural language descriptions. This capability is especially relevant for autonomous vehicles operating in unpredictable environments where new traffic signs or roadside objects may be encountered. The model's contribution lies in its semantic flexibility and capacity to generalize, paving the way for more robust and adaptive perception systems.

Chen et al. [6] developed a multi-view 3D object detection network tailored to fuse spatial information from multiple calibrated cameras. This architecture addresses key limitations of monocular systems, such as depth ambiguity and sensitivity to occlusion. By synthesizing views into a common spatial representation, the model improves detection accuracy in cluttered or partially obstructed scenes—conditions frequently encountered in dense urban traffic. The contribution of this work is in establishing an efficient and scalable multi-view fusion pipeline that balances geometric rigor with computational feasibility.

The BEVFormer series—comprising contributions from Li et al. [21], Yang et al. [54], and Pan et al. [31]—advances Bird's-Eye View (BEV) perception by transforming multi-view image data into a consistent top-down spatial representation. BEVFormer employs spatiotemporal transformers to integrate sequential observations, enhancing the temporal coherence of perceived scenes. Yang et al. refine this with modern visual backbones and perspective supervision, while Pan et al. enrich the system with CLIP-derived semantic embeddings for language-guided perception. These methods solve the problem of reconciling temporal, geometric, and semantic information into a unified spatial framework essential for planning and prediction in AVs.

Complementing these architectural developments, a set of comprehensive survey works—Arnold et al. [1], Qian et al. [33], Mao et al. [25], Feng et al. [10], and Feng et al. [11]—provides a critical synthesis of 3D perception for autonomous driving. These surveys chart the evolution of detection algorithms from traditional geometric methods to probabilistic and deep learning-based models, highlighting challenges such as sensor fusion, uncertainty modeling, and interpretability. Their significance lies in benchmarking research progress and identifying open questions, thereby guiding the design of more effective AV perception stacks.

5.2.3 Tracking and Temporal Association

Detection systems that analyze each video frame independently may yield inconsistent or unstable results, especially in real-world driving conditions where signs can be partially occluded, temporarily obscured, or outside the field of view for a few frames. Temporal tracking and association mechanisms solve this problem by linking the detected instances of traffic signs across consecutive frames. The goal is to maintain a continuous and coherent representation of each sign in the environment, even as its appearance changes slightly due to camera motion, viewing angle, or environmental factors.

The most common method used in traffic sign tracking is the Kalman filter, a recursive algorithm that estimates the current state of an object (in this case, the sign's position and velocity) based on a prior estimate and new observations. This method assumes a linear motion model and Gaussian noise, allowing it to predict where the sign will appear in the next frame. When the sign is detected again, the filter corrects its prediction using the new data, leading to more stable and accurate tracking over time. However, the Kalman filter's assumptions about linearity and Gaussian noise make it less effective in scenarios with non-linear dynamics or more erratic motion patterns.

To overcome these limitations, more advanced techniques like particle filters are employed. Particle filters maintain a set of possible hypotheses (particles) for the sign's position and weight them based on how well they explain the observations. This allows the system to track signs even in highly uncertain or cluttered environments. However, particle filters can be computationally expensive, which may pose challenges for real-time implementation.

Recently, data-driven tracking algorithms such as Deep SORT (Simple Online and Realtime Tracking with a deep appearance descriptor) have been adopted in autonomous systems. Deep SORT combines object detection with appearance-based re-identification features extracted from deep neural networks. It uses both the spatial location (from motion models) and appearance (from convolutional features) to maintain identity consistency across frames. This makes it particularly robust in handling occlusions, sign reappearance, and multiple overlapping signs.

Overall, tracking and temporal association not only enhance the stability of traffic sign detection but also reduce false negatives and provide essential temporal continuity. This consistency is crucial for downstream components of the AV pipeline, such as decision-making systems and map matching algorithms, which rely on persistent environmental awareness for safe and context-aware navigation.

TrackingNet/LaSOT Muller et al. and Fan et al. presented two robust benchmarks for tracking in unconstrained environments [29, 9].

Online and Real-time MOT Wang et al. advanced low-latency tracking algorithms suitable for AV systems [48].

Kernel Tracking Comaniciu et al. established a widely used kernel-based object tracker [8].

Fast Segmentation and Tracking Wang et al. proposed a unified method to track and segment objects on-the-fly [46].

5.2.4 Challenges and Robustness

Anomaly Detection Bogdoll et al. surveyed detection of unusual objects and behaviors in AV environments [3].

Radar Perception Zhou et al. reviewed deep learning methods and datasets for radar signal processing [63].

Foundational Perception Studies Aufrère et al. and Gruyer et al. explored perception strategies for collision avoidance and decision-making [2, 14].

Waymo Open Dataset Sun et al. released a large-scale multimodal perception dataset for urban driving [42].

DeepDriving Chen et al. proposed a deep learning-based affordance model for direct perception control [5].

Energy-Efficient Middleware Liu et al. introduced E2M for optimized edge inference in mobile robotics [23].

Surveys on Learning-Based AVs Grigorescu et al. and Kiran et al. synthesized research on deep learning and reinforcement learning in AV systems [13, 19].

Despite technological advancements, traffic sign detection and tracking remain challenging due to the diverse and dynamic conditions encountered in real-world driving environments. One of the primary obstacles is visual occlusion, where signs are partially or fully blocked by other vehicles, pedestrians, roadside objects, or environmental elements such as overgrown trees. These occlusions may be transient, but they can prevent consistent detection if the perception system relies only on instantaneous frame-by-frame analysis.

Another critical challenge arises from the variability in lighting conditions. Bright sunlight can produce glare or harsh shadows that obscure the sign's features, while overcast skies or nighttime driving reduce overall contrast, making it difficult to distinguish signs from the background. Headlight reflections, rain droplets on the camera lens, and artificial lighting sources can further distort visual input. These lighting issues are especially problematic for color-dependent detection methods and may necessitate the use of alternative sensing modalities or more robust feature representations.

Environmental wear and tear significantly affect the legibility of traffic signs. Exposure to the elements over time can cause signs to fade, peel, rust, or become dirty, altering their appearance in ways that deviate from the training data. Construction zones introduce temporary signage that may not conform to standard formats, while vandalism or misalignment can mislead the detection system. These degradations necessitate a high degree of generalization from detection algorithms, which must recognize the semantic content of a sign even when its visual form is impaired.

In addition to physical and visual degradation, contextual relevance poses a unique challenge. AVs often operate in multi-lane roads or complex interchanges where multiple signs may be visible simultaneously. Not all visible signs are applicable to the vehicle's current route or lane. For example, an exit sign may be visible from a highway lane but is only relevant to a merging vehicle. Misinterpreting such signs can lead to unsafe or inefficient driving behavior. Therefore, sign detection must be contextualized using spatial reasoning and localization data.

Another increasing concern involves the detection of complex, composite, or unconventional traffic signs. In many urban environments, drivers may encounter signs that contain multiple instructions or are presented in an unstructured format, such as stacked signs that combine speed limits with construction warnings and temporary detour directions. These composite signs can be difficult for detection systems to parse correctly, particularly when they contain conflicting instructions or require conditional reasoning, such as time-of-day applicability. Systems must not only detect such signs but also segment and interpret each component accurately in order to inform downstream behavior planning.

Additionally, autonomous vehicles may need to interpret hand-written or improvised traffic notices, such as those used at construction sites or during temporary events. These notices are often affixed to poles or barricades, written on cardboard or paper, and lack the standardized symbology that conventional traffic signs exhibit. They may use ambiguous language, non-standard fonts, or be partially obscured or poorly placed. Detecting and interpreting such signs poses a significant challenge to current deep learning systems, which are typically trained on curated datasets featuring high-quality, standardized signs. Addressing this issue may require incorporating scene text recognition (OCR) pipelines, pre-trained on diverse fonts and writing styles, along with contextual reasoning modules that consider the overall traffic scene to assess the plausibility and intent of such notices.

To address these challenges, robust training procedures are essential. Data augmentation is a widely used technique to synthetically replicate challenging conditions during training. This includes applying geometric transformations (e.g., rotation, scaling), photometric changes (e.g., brightness, contrast, noise), and occlusion simulations to improve the network's ability to generalize. Additionally, multi-sensor fusion—integrating data from LiDAR, radar, and GPS—can provide complementary information to validate or correct camera-based detections.

Finally, the integration of high-definition (HD) maps and scene understanding modules allows the AV to verify the plausibility of detected signs within a geospatial context. For instance, if a stop sign is detected on a highway where no such sign exists in the map database, the system can flag the detection as potentially erroneous. Combining vision with spatial and semantic context is critical for achieving robust and accurate perception in the face of these diverse and complex challenges.

5.2.5 Lane Detection and Tracking

Lane detection and tracking is a critical component of the perception system in autonomous vehicles (AVs), enabling the vehicle to understand its lateral position on the road and to follow traffic rules related to lane discipline. This capability underpins numerous driving functions such as lane keeping, lane changing, and merge maneuvers, and it must remain reliable under a wide range of visual and environmental conditions. Accurate lane detection ensures that the AV maintains an appropriate trajectory within its lane, preventing unintended departures and supporting transitions such as exits, merges, and turns. Furthermore, high-precision lane detection is foundational for planning modules that depend on lane-level geometry to generate safe and context-aware behaviors.

Traditional lane detection techniques rely on classical computer vision pipelines structured around interpretable, low-level image features—such as edges, gradients, and contours—to infer the presence of lane markings. These pipelines typically begin with image preprocessing steps, such as Gaussian blurring, to reduce noise and enhance the signal-to-noise ratio. This is followed by edge detection methods like the Sobel or Canny operators, which identify gradient discontinuities corresponding to lane boundaries.

A key insight is that lane markings usually appear as elongated, high-contrast lines extending vertically from the bottom of a forward-facing image. The Hough Transform is employed to detect straight-line segments by transforming edge data into a parametric representation. This approach is particularly effective for identifying linear lanes on highways, even when partially occluded or degraded. To further simplify the geometric interpretation of lanes, perspective transforms are used to simulate a bird's-eye view (BEV) of the road. In this transformed space, lane lines appear parallel and equidistant, making them easier to process using rule-based or symmetry-based heuristics.

However, the real-world performance of classical methods is often compromised by variations in lighting, occlusion, road surface quality, and inconsistent or missing lane markings. Their rigid assumptions about geometry and contrast reduce their robustness under diverse operational conditions, limiting their deployment to well-maintained, high-visibility roads.

To overcome these limitations, deep learning-based approaches have emerged that treat lane detection as a semantic segmentation problem. These systems leverage convolutional neural networks (CNNs) to assign a lane/non-lane label to each pixel, enabling detection even when lane markings are faded or occluded. Spatial CNNs like SCNN extend this approach by introducing directional message passing, improving continuity along the orientation of elongated structures such as lanes. This mechanism allows the network to better understand lane geometry and infer missing sections through context propagation.

Wang et al. [47] introduced LaneNet, a real-time, deep learning-based lane detection system designed to segment lane markings and distinguish individual lane instances. The architecture comprises two main branches: a binary segmentation branch that separates lane from non-lane pixels, and an embedding branch that maps pixels to an embedding space to enable instance-level differentiation. Each detected lane line is encoded as a distinct cluster in the embedding space, allowing the model to differentiate between closely spaced or parallel lanes—a critical need for multi-lane roads and highway settings. LaneNet uses an encoder-decoder structure based on lightweight CNN blocks, which ensures computational efficiency for deployment in embedded AV platforms. Additionally, it introduces a custom loss function that enforces intra-cluster compactness and inter-cluster separation in the embedding space, improving performance in complex visual environments with occlusions, shadowing, or degraded lane markings. LaneNet's balance of speed, segmentation accuracy, and instance awareness has made it a foundational model for AV lane perception tasks.

Comprehensive reviews by Zakaria et al., Lee and Liu, and Chetan et al. [59, 20, 7] have examined lane detection techniques with respect to accuracy, temporal consistency, and robustness to environmental changes. These studies offer comparative evaluations and outline key challenges, including the generalization of models to unseen road configurations and the trade-offs between computational cost and detection precision.

Ros et al. and Thorpe et al. [38, 44] laid the groundwork for modern lane detection with early vision-based systems. These early efforts integrated lane detection with control systems and helped define the architecture of AV perception pipelines, particularly by validating the feasibility of visual-only road understanding in structured driving environments.

Together, these methods—from classic image processing to deep parametric inference—illustrate the evolution of lane detection into a mature field that integrates geometric reasoning, visual learning, and real-time constraints. Lane detection remains one of the most mission-critical elements in AV perception, as it informs motion planning and ensures regulatory compliance in structured road environments.

5.2.6 *Training and Dataset Diversity*

Both models benefit from training on datasets like CULane[32] and BDD100K[58], which present diverse scenarios including multi-lane highways, urban intersections, varying weather conditions, and different lighting situations. The variety in these datasets ensures that the models generalize across road types and can robustly infer lane boundaries even when partial occlusions, inconsistent markings, or complex road topologies are present. Their success illustrates how embedding task-specific structure into the model architecture can improve both accuracy and operational feasibility.

Tracking builds upon detection by ensuring that identified lane boundaries remain coherent and consistent over time. This temporal stability is essential for maintaining vehicle control and planning under conditions where detections may intermittently fail. Kalman filters are widely used for this purpose, estimating lane positions and curvatures based on previous states and updating them using new observations. These filters assume linear or polynomial motion models and help smooth the vehicle's perception of lane boundaries, especially when some frames are unreliable. Polynomial fitting is often used in tandem to model the shape of lanes across a temporal window, fitting parabolas or splines to recent detections to account for road curvature.

In more complex scenarios involving multiple interpretations of road layout—such as forks, lane merges, or construction detours—tracking may use particle filters or multiple hypothesis tracking (MHT). These techniques maintain a set of candidate lane hypotheses and evaluate them against incoming data to identify the most probable interpretation. Integration with IMUs and GPS further refines tracking by incorporating inertial and positional cues, allowing the AV to estimate lane-relative pose even when visual cues are weak or absent.

Despite these advancements, lane detection and tracking systems face significant challenges in real-world conditions. Variability in weather such as heavy rain, fog, and snow reduces visibility and often obscures lane markings. The reflective properties of water on road surfaces can produce misleading edges or eliminate contrast altogether. Road construction introduces temporary markings and inconsistent signage, while human activities—such as repainting or obstructing markings—introduce further noise. The inconsistency in marking standards across regions and countries adds complexity to generalizing any lane detection model.

Moreover, AVs are increasingly expected to operate in unstructured or semi-structured environments where explicit lane boundaries are missing. Examples include parking lots, warehouse yards, and rural dirt roads. In these cases, the AV must infer drivable space from context cues such as curbs, vehicle tracks, or even pedestrian paths. This requires integrating lane detection with drivable area segmentation, which often involves fusing RGB camera data with depth information from LiDAR or radar. These hybrid systems call for advanced sensor fusion techniques and dynamic context understanding to ensure safe navigation.

5.3 Anomaly Detection and Mitigation

Anomaly detection and correction are essential components of perception in autonomous vehicles (AVs), enabling the system to identify and adapt to unexpected deviations in sensor inputs or inferred environmental features. These deviations may signal critical errors, potential hazards, or out-of-distribution (OOD) conditions. Unlike in controlled test environments, real-world driving presents AVs with an unbounded set of rare, edge-case, or previously unseen scenarios that cannot be comprehensively captured during initial training or calibration. These include abrupt weather shifts, erratic pedestrian behaviors, sensor interference from reflective or absorptive surfaces, or novel traffic configurations.

In the absence of anomaly detection, an AV assumes that all input data and model inferences are valid, making it vulnerable to misinterpretation and unsafe behavior. For example, a misclassified object may be dismissed as background when it represents an imminent hazard, or a misread lane marking may lead to improper steering corrections. Anomaly detection acts as a critical self-awareness mechanism, flagging data inconsistencies or uncertain decisions that deviate from established norms. This allows the system to activate cautionary responses, revert to safe fallback behaviors, or transfer control to a human operator.

Complementing detection, anomaly correction plays an equally important role by enabling the system to respond effectively once anomalies are identified. This may involve filtering corrupted data, estimating missing values, or drawing on sensor redundancy to validate observations. Together, detection and correction form a resilience layer that bolsters perception not just under ideal conditions but also amidst the uncertainties and complexities of real-world environments. Such anomalies may stem from hardware faults, sensor occlusions, adversarial noise, or novel driving situations unobserved during training. Rapid identification and mitigation of these irregularities is essential to maintaining AV reliability and safety.

At a foundational level, anomaly detection seeks to distinguish between normal and abnormal patterns in sensory input. Classical statistical approaches address this by modeling the distribution of typical features and identifying deviations. Techniques such as Gaussian mixture models (GMMs), principal component analysis (PCA), and Mahalanobis distance measures characterize the expected structure of scene features or trajectories. These methods are computationally efficient and relatively interpretable but often underperform in high-dimensional or temporally complex data typical of AV environments.

Deep learning has introduced more flexible and expressive frameworks for anomaly detection by leveraging the representational power of neural networks. Autoencoders are commonly used in this context: trained to reconstruct normal input data, they flag inputs as anomalous when reconstruction errors exceed a learned threshold. Variational autoencoders (VAEs) and generative adversarial networks (GANs) extend this framework by learning latent distributions or probabilistic constraints, allowing finer sensitivity to subtle deviations.

Uncertainty-aware inference provides another strategy. Techniques such as Monte Carlo dropout, deep ensembles, and Bayesian neural networks estimate predictive confidence or variance. Outputs with low confidence or high variance suggest unreliable inferences, which can be interpreted as potential anomalies. This is especially valuable in high-stakes modules like object detection or segmentation, where misclassifying unknown classes or ambiguous inputs can lead to unsafe behavior. Uncertainty estimates can trigger cautionary responses or solicit additional validation from alternate sensors.

Beyond detection, correction strategies aim to mitigate the effects of flagged anomalies. Sensor fusion plays a central role here, as combining data from cameras, LiDAR, and radar allows cross-modality validation. When one sensor is degraded, others can compensate, reducing the risk of perception failures. Additional techniques include temporal smoothing, predictive filtering, and model-based interpolation to fill in gaps or correct outliers. Rule-based overrides and behavior arbitration modules can enforce safe fallback maneuvers when confidence thresholds are violated.

Emerging research in self-supervised and contrastive learning offers promising avenues for anomaly detection. These models are trained to predict future sensor states, scene evolution, or cross-modal alignments without explicit supervision. Deviations from these learned predictions indicate anomalies not only in visual space but also in semantic and temporal structures. This broadens detection capabilities and facilitates generalization to new environments.

Bogdoll et al. [4] provide a comprehensive survey of datasets available for evaluating perception-based anomaly detection in autonomous driving. Their study reveals a notable gap in current benchmarks, which tend to favor clean, well-labeled, and typical scenes. They emphasize the need for dedicated datasets that include rare events, environmental perturbations, and sensor faults to reflect real-world operational diversity. The survey categorizes existing datasets by anomaly types and highlights missing evaluation protocols, offering valuable guidance for future dataset curation and model testing.

Ultimately, integrating robust anomaly detection and correction mechanisms into the AV perception stack enhances the system's ability to respond intelligently to uncertainty, reducing the likelihood of cascading errors. As autonomous systems expand into increasingly dynamic and diverse deployment contexts, these capabilities will be essential for achieving resilient and trustworthy long-term autonomy.

5.4 Scene Understanding in Autonomous Vehicle Perception

Scene understanding is a high-level cognitive function within the perception stack of autonomous vehicles (AVs) that enables the interpretation of complex environments. It integrates information from multiple low-level perception outputs—such as object detection, semantic segmentation, lane detection, and depth estimation—into a coherent representation of the scene. This structured understanding is crucial not only for tracking and recognizing individual elements but also for inferring spatial, temporal, and social relationships among dynamic agents and environmental features.

The role of scene understanding is to move beyond recognition and toward comprehension. While object detectors may identify a pedestrian, scene understanding places that pedestrian in context—on a crosswalk, waiting to cross, or moving toward a vehicle's path. It incorporates semantic and geometric data, motion patterns, and prior map knowledge to infer both the state and potential behavior of objects in the scene. This predictive capability is vital for safe and anticipatory decision-making in urban and highway driving.

In the AV perception pipeline, scene understanding provides the situational awareness necessary for robust and adaptive behavior. It connects perception with planning by constructing a dynamic model of the world that evolves over time. This involves multi-object tracking, agent interaction modeling, road structure estimation, and event inference. Through temporal continuity, the system can recognize activities such as a car preparing to turn, a cyclist weaving through traffic, or a pedestrian hesitating before stepping into the road.

Technically, scene understanding draws on spatiotemporal deep learning methods such as graph neural networks, recurrent neural networks, and attention mechanisms. These tools allow the vehicle to reason over relationships between agents and across time, supporting intent estimation and long-horizon forecasting. Moreover, fusion with HD maps and routing information helps refine the understanding of traffic rules and lane structures within the scene.

Scene understanding differs from other perception tasks in its emphasis on relational and predictive modeling. Whereas traditional vision modules operate on independent detections or segmentations, scene understanding binds these elements into a structured semantic graph or world model. This abstraction layer enables planning systems to reason at a higher level of granularity, incorporating intent and context rather than just position and classification.

Ultimately, scene understanding is fundamental to the safe deployment of AVs. It ensures that perception is not merely reactive but proactively aware of potential risks and socially intelligent behaviors. As autonomous systems operate in increasingly complex and interactive environments, the quality and depth of their scene understanding will define their ability to navigate safely and effectively.

In addition to the architectures described in the following sections, foundational contributions to this field include early surveys on scene understanding (e.g., Yang et al. [55]), multimodal deep learning frameworks (Huang et al. [17]), and synthetic-to-real domain adaptation via spatial CNNs (Pan et al. [32]). Large-scale pretraining paradigms such as DriveWorld (Min et al. [27]) further demonstrate the importance of leveraging world models for pretext tasks to enrich semantic scene representations. Comprehensive reviews by Guo et al. [15] and Muhammad et al. [28] outline the evolution of scene understanding from modular stacks to unified end-to-end learning systems, identifying the technical and deployment challenges that still persist.

5.4.1 State-of-the-Art Approaches in Scene Understanding

Recent advancements in scene understanding for autonomous vehicles have moved decisively toward end-to-end, multimodal, and semantically rich models that tightly integrate perception, reasoning, and decision-making. Among the most representative developments is the EMMA family of models.

Hwang et al. introduced EMMA [18], a comprehensive end-to-end framework that fuses camera, LiDAR, and HD-map inputs through multi-branch fusion and transformer-based encoding. EMMA jointly optimizes perception, prediction, and planning, thereby modeling agent interactions over time and supporting complex, context-aware decision-making.

Building on EMMA's foundation, Xing et al. released OpenEMMA [53], an open-source platform designed to increase reproducibility and transparency. By offering standardized datasets, pre-trained weights, and modular APIs, OpenEMMA facilitates benchmarking and community-driven development.

Qiao et al. then proposed LightEMMA [34], a resource-efficient derivative that minimizes inference latency via compressed attention modules and compact model design. Its relevance lies in enabling deployment on edge devices without compromising accuracy—an essential step toward real-time, cost-efficient AV systems.

In parallel, Zhao et al. developed Sce2DriveX [62], a multimodal large language model (MLLM) framework that treats scene understanding as a language-based reasoning task. Sce2DriveX maps structured perceptual inputs into tokenized prompts, allowing pre-trained VLMs to process traffic scenarios and generate interpretable driving actions. This approach introduces semantic generalization and open-world capability, particularly important for diverse and unstructured environments.

Focusing on relational inference, Song et al. proposed InsightDrive [40], which learns scene graphs that encode spatial constraints and agent intent. The system produces graph-based embeddings for downstream planning, making predictions more interpretable and traceable. InsightDrive contributes significantly to safe operation in multi-agent contexts and aligns with safety-critical demands for explainable reasoning.

Expanding the sensor domain, Wu et al. presented TARS [50], a radar-focused model designed for dense scene flow estimation in degraded visibility. TARS estimates object motion directly from radar, mitigating the limitations of vision under fog, rain, or occlusion. Its use of specialized filtering for radar granularity introduces robustness vital for all-weather autonomy.

Vision-language models (VLMs) form another pillar of recent innovation. Zhang et al. [60] provide an architectural overview of how VLMs can unify language and perception for semantic segmentation and intent grounding. Liu et al. [22] complement this by analyzing hallucination and grounding failures, offering classification schemes and mitigation strategies for integrating LLMs in safety-critical AV perception.

Pioneering contributions such as CLIP [35] have enabled alignment between image embeddings and textual labels, laying the foundation for open-vocabulary scene understanding. The Segment Anything Model (SAM) [64] advances this further by offering a universal segmentation interface, adaptable to novel categories and prompts with minimal supervision.

5.5 Hands-on Exercises

5.5.1 *Exercise 1: Object Detection with YOLO*

Task Develop a real-time object detection pipeline using a pre-trained YOLOv5 or YOLOv8 model.

Objective Download a traffic dataset (e.g., BDD100K or KITTI), perform inference to visualize bounding boxes, and tune detection parameters (confidence threshold, non-max suppression) for performance evaluation.

Deliverable Submit a short report including:

- Screenshots of detection results on sample images.
- Plots of precision and recall at various thresholds.
- Optional: Annotated examples of false positives and their interpretation.

5.5.2 *Exercise 2: Lane Detection with LaneNet*

Task Train and evaluate a lane detection model using LaneNet to segment lanes from front-facing video.

Objective Use TuSimple or CULane datasets. Train both binary segmentation and instance embedding branches. Optionally apply a perspective transform for bird's-eye view projection.

Deliverable Provide:

- Evaluation metrics (e.g., IoU, accuracy).
- Visualizations of segmented lane masks.
- Optional: BEV-rendered lane projection.

5.5.3 *Exercise 3: Multi-Object Tracking with Deep SORT*

Task Implement a multi-object tracking system using Deep SORT combined with a YOLO detector.

Objective Apply object detection on video frames and associate detections over time using Deep SORT. Tune re-identification and appearance parameters for robust ID tracking.

Deliverable Submit:

- A video showing tracked bounding boxes with stable IDs.
- Code or log output demonstrating tracking through occlusion.

5.5.4 *Exercise 4: Scene Segmentation with Segment Anything*

Task Apply prompt-based segmentation to a traffic scene using the Segment Anything Model (SAM).

Objective Use images from a traffic dataset and apply SAM prompts (click, box, text) to segment regions of interest (e.g., vehicles, signs, pedestrians). Compare with manual ground truth if available.

Deliverable Include:

- Segment masks produced by different prompt types.
- Qualitative comparison with manual annotation or baseline segmentation.

5.5.5 *Exercise 5: Scene Graph Planning with InsightDrive*

Task Use structured scene representations to drive simple rule-based planning.

Objective Convert object detection outputs into scene graphs encoding spatial relations. Visualize graph evolution over time and implement a rule-based planner reacting to specific configurations (e.g., pedestrian ahead, vehicle cutting in).

Deliverable Submit:

- Sample scene graph visualizations.
- A short video or simulation trace showing planned behavior changes in response to graph updates.

5.5.6 *Exercise 6: Anomaly Detection in Perception Pipelines*

Task Simulate and detect anomalies in visual inputs using learned or statistical models.

Objective Introduce noise, occlusion, or out-of-distribution frames. Apply an autoencoder or Mahalanobis-distance-based scoring method to identify anomalous inputs.

Deliverable Provide:

- Example inputs with anomalies and corresponding anomaly heatmaps or scores.
- ROC curve or detection precision analysis.

Recommended Papers to Read

1. Xing, S., Qian, C., Wang, Y., Hua, H., Tian, K., Zhou, Y., & Tu, Z. (2025). Openemma: Open-source multimodal model for end-to-end autonomous driving. In Proceedings of the Winter Conference on Applications of Computer Vision (pp. 1001–1009). [53]
2. Zhao, R., Yuan, Q., Li, J., Hu, H., Li, Y., Zheng, C., & Gao, F. (2025). Sce2DriveX: A generalized MLLM framework for scene-to-drive learning. arXiv preprint arXiv:2502.14917. [62]
3. Guo, Z., Huang, Y., Hu, X., Wei, H., & Zhao, B. (2021). A survey on deep learning based approaches for scene understanding in autonomous driving. Electronics, 10(4), 471. [15]

References

1. Arnold, E., et al. (2019). A survey on 3d object detection methods for autonomous driving applications. *IEEE Transactions on Intelligent Transportation Systems*, *20*(10), 3782–3795.
2. Aufrère, R., et al. (2003). Perception for collision avoidance and autonomous driving. *Mechatronics*, *13*(10), 1149–1161.
3. Bogdoll, D., Nitsche, M., & Zöllner, J. M. (2022). Anomaly detection in autonomous driving: A survey. In *Proceedings of the IEEE/CVF Conference on Computer Vision and Pattern Recognition* (pp. 4488–4499).
4. Bogdoll, D., et al. (2023). Perception datasets for anomaly detection in autonomous driving: A survey. In *2023 IEEE Intelligent Vehicles Symposium (IV)* (pp. 1–8). IEEE.
5. Chen, C., et al. (2015). Deepdriving: Learning affordance for direct perception in autonomous driving. In *Proceedings of the IEEE International Conference on Computer Vision* (pp. 2722–2730).
6. Chen, X., et al. (2017). Multi-view 3d object detection network for autonomous driving. In *Proceedings of the IEEE Conference on Computer Vision and Pattern Recognition* (pp. 1907–1915).
7. Chetan, N. B., et al. (2020). An overview of recent progress of lane detection for autonomous driving. In *2019 6th International Conference on Dependable Systems and Their Applications (DSA)* (pp. 341–346). IEEE.
8. Comaniciu, D., Ramesh, V., & Meer, P. (2003). Kernel-based object tracking. *IEEE Transactions on Pattern Analysis and Machine Intelligence*, 25(5), 564–577.

9. Fan, H., et al. (2019). LaSOT: A high-quality benchmark for large-scale single object tracking. In *Proceedings of the IEEE/CVF Conference on Computer Vision and Pattern Recognition* (pp. 5374–5383).
10. Feng, D., et al. (2020). Deep multi-modal object detection and semantic segmentation for autonomous driving: Datasets, methods, and challenges. *IEEE Transactions on Intelligent Transportation Systems, 22*(3), 1341–1360.
11. Feng, D., et al. (2021). A review and comparative study on probabilistic object detection in autonomous driving. *IEEE Transactions on Intelligent Transportation Systems, 23*(8), 9961–9980.
12. Ge, Z., et al. (2021). *Yolox: Exceeding yolo series in 2021*. Preprint. arXiv:2107.08430.
13. Grigorescu, S., et al. (2020). A survey of deep learning techniques for autonomous driving. *Journal of Field Robotics, 37*(3), 362–386.
14. Gruyer, D., et al. (2017). Perception, information processing and modeling: Critical stages for autonomous driving applications. *Annual Reviews in Control, 44*, 323–341.
15. Guo, Z., et al. (2021). A survey on deep learning based approaches for scene understanding in autonomous driving. *Electronics, 10*(4), 471.
16. Houben, S., et al. (2013). Detection of traffic signs in real-world images: The German Traffic Sign Detection Benchmark. In *The 2013 International Joint Conference on Neural Networks (IJCNN)* (pp. 1–8). IEEE.
17. Huang, Z., et al. (2020). Multi-modal sensor fusion-based deep neural network for end-to-end autonomous driving with scene understanding. *IEEE Sensors Journal, 21*(10), 11781–11790.
18. Hwang, J.-J., et al. (2024). *EMMA: End-to-end multimodal model for autonomous driving*. Preprint. arXiv:2410.23262.
19. Kiran, B. R., et al. (2021). Deep reinforcement learning for autonomous driving: A survey. *IEEE Transactions on Intelligent Transportation Systems, 23*(6), 4909–4926.
20. Lee, D.-H., & Liu, J.-L. (2023). End-to-end deep learning of lane detection and path prediction for real-time autonomous driving. *Signal, Image and Video Processing, 17*(1), 199–205.
21. Li, Z., et al. (2024). Bevformer: learning bird's-eye-view representation from lidar-camera via spatiotemporal transformers. *IEEE Transactions on Pattern Analysis and Machine Intelligence, 47*, 2020–2036.
22. Liu, H., et al. (2024). *A survey on hallucination in large vision-language models*. Preprint. arXiv:2402.00253.
23. Liu, L., et al. (2019). E2M: An energy-efficient middleware for computer vision applications on autonomous mobile robots. In *Proceedings of the 4th ACM/IEEE Symposium on Edge Computing* (pp. 59–73).
24. Luo, W., et al. (2021). Multiple object tracking: A literature review. *Artificial Intelligence, 293*, 103448.
25. Mao, J., et al. (2023). 3D object detection for autonomous driving: A comprehensive survey. *International Journal of Computer Vision, 131*(8), 1909–1963.
26. Meinhardt, T., et al. (2022). Trackformer: Multi-object tracking with transformers. In *Proceedings of the IEEE/CVF Conference on Computer Vision and Pattern Recognition* (pp. 8844–8854).
27. Min, C., et al. (2024). Driveworld: 4d pre-trained scene understanding via world models for autonomous driving. In *Proceedings of the IEEE/CVF Conference on Computer Vision and Pattern Recognition* (pp. 15522–15533).
28. Muhammad, K., et al. (2022). Vision-based semantic segmentation in scene understanding for autonomous driving: Recent achievements, challenges, and outlooks. *IEEE Transactions on Intelligent Transportation Systems, 23*(12), 22694–22715.
29. Muller, M., et al. (2018). Trackingnet: A large-scale dataset and benchmark for object tracking in the wild. In *Proceedings of the European conference on computer vision (ECCV)* (pp. 300–317).
30. O'shea, K., & Nash, R. (2015). *An introduction to convolutional neural networks*. Preprint. arXiv:1511.08458.

31. Pan, C., et al. (2024). CLIP-BEVFormer: Enhancing multi-view image-based BEV detector with ground truth flow. In *Proceedings of the IEEE/CVF Conference on Computer Vision and Pattern Recognition* (pp. 15216–15225).
32. Pan, X., et al. (2018). Spatial as deep: Spatial CNN for traffic scene understanding. In *Proceedings of the AAAI Conference on Artificial Intelligence* (Vol. 32, Chap. 1).
33. Qian, R., Lai, X., & Li, X. (2022). 3D object detection for autonomous driving: A survey. *Pattern Recognition*, *130*, 108796.
34. Qiao, Z., et al. (2025). LightEMMA: Lightweight end-to-end multimodal model for autonomous driving. Preprint. arXiv:2505.00284.
35. Radford, A., et al. (2021). Learning transferable visual models from natural language supervision. In *International Conference on Machine Learning*. PmLR (pp. 8748–8763).
36. Redmon, J., et al. (2016). You only look once: Unified, real-time object detection. In *Proceedings of the IEEE Conference on Computer Vision and Pattern Recognition* (pp. 779–788).
37. Ren, S., et al. (2016). Faster R-CNN: Towards real-time object detection with region proposal networks. *IEEE Transactions on Pattern Analysis and Machine Intelligence*, *39*(6), 1137–1149.
38. Ros, G., et al. (2015). Vision-based offline-online perception paradigm for autonomous driving. In *2015 IEEE Winter Conference on Applications of Computer Vision* (pp. 231–238). IEEE.
39. Shi, H., Dao, S. D., & Cai, J. (2025). LLMFormer: large language model for open-vocabulary semantic segmentation. *International Journal of Computer Vision*, *133*(2), 742–759.
40. Song, R., et al. (2025). *InsightDrive: Insight scene representation for end-to-end autonomous driving*. Preprint. arXiv:2503.13047.
41. Soylu, E., & Soylu, T. (2024). A performance comparison of YOLOv8 models for traffic sign detection in the Robotaxi-full scale autonomous vehicle competition. *Multimedia Tools and Applications*, *83*(8), 25005–25035.
42. Sun, P., et al. (2020). Scalability in perception for autonomous driving: Waymo open dataset. In *Proceedings of the IEEE/CVF Conference on Computer Vision and Pattern Recognition* (pp. 2446–2454).
43. Sun, P., et al. (2020). *Transtrack: Multiple object tracking with transformer*. Preprint. arXiv:2012.15460.
44. Thorpe, C., et al. (1991). Toward autonomous driving: The CMU Navlab. I. Perception. *IEEE Expert*, *6*(4), 31–42.
45. Wang, J., et al. (2023). Improved YOLOv5 network for real-time multi-scale traffic sign detection. *Neural Computing and Applications*, *35*(10), 7853–7865.
46. Wang, Q., et al. (2019). Fast online object tracking and segmentation: A unifying approach. In *Proceedings of the IEEE/CVF conference on Computer Vision and Pattern Recognition* (pp. 1328–1338).
47. Wang, Z., Ren, W., & Qiu, Q. (2018). *Lanenet: Real-time lane detection networks for autonomous driving*. Preprint. arXiv:1807.01726.
48. Wang, Z., et al. (2020). Towards real-time multi-object tracking. In *European Conference on Computer Vision* (pp. 107–122). Springer.
49. Wojke, N., Bewley, A., & Paulus, D. (2017). Simple online and realtime tracking with a deep association metric. In *2017 IEEE International Conference on Image Processing (ICIP)* (pp. 3645–3649). IEEE.
50. Wu, J., et al. (2025). *TARS: Traffic-aware radar scene flow estimation*. Preprint. arXiv:2503.10210.
51. Wu, Y., Lim, J., & Yang, M.-H. (2013). Online object tracking: A benchmark. In *Proceedings of the IEEE Conference on Computer Vision and Pattern Recognition* (pp. 2411–2418).
52. Xiang, Y. Alahi, A., & Savarese, S. (2015). Learning to track: Online multi-object tracking by decision making. In *Proceedings of the IEEE International Conference on Computer Vision* (pp. 4705–4713).
53. Xing, S., et al. (2025). Openemma: Open-source multimodal model for end-to-end autonomous driving. In *Proceedings of the Winter Conference on Applications of Computer Vision* (pp. 1001–1009).

54. Yang, C., et al. (2023). BEVFormer v2: Adapting modern image backbones to bird's-eye-view recognition via perspective supervision. In *Proceedings of the IEEE/CVF Conference on Computer Vision and Pattern Recognition* (pp. 17830–17839).
55. Yang, S., et al. (2018). Scene understanding in deep learning-based end-to-end controllers for autonomous vehicles. *IEEE Transactions on Systems, Man, and Cybernetics: Systems, 49*(1), 53–63.
56. Yang, Y., et al. (2015). Towards real-time traffic sign detection and classification. *IEEE Transactions on Intelligent Transportation Systems, 17*(7), 2022–2031.
57. Yilmaz, A., Javed, O., & Shah, M. (2006). Object tracking: A survey. *ACM Computing Surveys (CSUR), 38*(4), 13–es.
58. Yu, F., et al. (2020). BDD100K: A diverse driving dataset for heterogeneous multitask learning. In *Proceedings of the IEEE/CVF Conference on Computer Vision and Pattern Recognition* (pp. 2636–2645).
59. Zakaria, N. J., et al. (2023). Lane detection in autonomous vehicles: A systematic review. *IEEE Access, 11*, 3729–3765.
60. Zhang, J., et al. (2024). Vision-language models for vision tasks: A survey. *IEEE Transactions on Pattern Analysis and Machine Intelligence, 46*(8), 5625–5644.
61. Zhang, L., et al. (2025). TSD-DETR: A lightweight real-time detection transformer of traffic sign detection for long-range perception of autonomous driving. *Engineering Applications of Artificial Intelligence, 139*, 109536.
62. Zhao, R., et al. (2025). *Sce2DriveX: A generalized MLLM framework for scene-to-drive learning*. Preprint. arXiv:2502.14917.
63. Zhou, Y., et al. (2022). Towards deep radar perception for autonomous driving: Datasets, methods, and challenges. *Sensors, 22*(11), 4208.
64. Zhou, Y., et al. (2023). Segment everything everywhere all at once. *Advances in Neural Information Processing Systems, 36*, 19769–19782.

Chapter 6
Localization Algorithms

6.1 Introduction

When people refer to *localization*, they typically mean obtaining a rough estimate of a target's position on a map. For example, Professor X might be located in his office on the 15th floor, or my current GPS coordinates might be 39.649548, −75.789969. For humans, precise positioning isn't always necessary for route planning. However, for autonomous vehicles, localization accuracy is critical. The system must determine not just a general location—like "somewhere on the Street"—but precise information such as the specific lane, the distance from the curb, and the vehicle's orientation. This level of detail is essential for the onboard computing system to ensure safe and reliable operation.

In a typical autonomous driving system, the localization module performs two primary tasks:

1. **High-frequency pose tracking**: Delivers real-time updates on the vehicle's position to support navigation and object detection.
2. **Low-frequency global re-localization**: Estimates a coarse position to reinitialize the tracking algorithm when pose tracking fails.

The implementation of the localization module is closely tied to the types of sensors installed on the vehicle. In this chapter, we begin by providing an overview of the localization problem, then discuss localization using four different sensor modalities, and finally present the latest research developments in the field.

Contributor: Ren Zhong

W. Shi, Y. He, *Introduction to Autonomous Driving*,
https://doi.org/10.1007/978-3-031-99485-2_6

6.2 Definition of Pose for Localization

Before discussing solutions for obtaining localization, it is important to first define what a pose is. A pose refers to the position and orientation of an object in space, typically represented as a combination of coordinates (x, y, z) and rotation in Euler angles (roll, pitch, yaw). It is always defined relative to a coordinate system. In a typical autonomous driving system, there are usually three of coordinate systems:

- **World frame (map)**—A fixed coordinate system aligned with the environment.
- **Vehicle frame**—A coordinate system attached to the vehicle.
- **Sensor frame**—Each sensor (camera, LiDAR, IMU) has its own local coordinate system.

After the localization system is initialized, the initial pose of the sensor frame is typically referred to as the odometry frame. Subsequent poses returned by the localization system are defined relative to this frame. As a result, it becomes necessary to transform poses between different coordinate systems. To facilitate this process, poses are usually represented using homogeneous transformation matrices.

$$P = \begin{bmatrix} R & \mathbf{t} \\ 0 & 1 \end{bmatrix}$$

Where:

- $R \in \mathbb{R}^{3\times3}$: a rotation matrix that represents orientation
- $\mathbf{t} \in \mathbb{R}^3$: a translation vector that represents position

Given the relative pose T between the world frame and the odometry frame, a pose $\mathbf{p}_{\text{sensor}}$ obtained by the sensor can be transformed into world coordinates as follows:

$$\mathbf{p}_{\text{world}} = T \cdot \mathbf{p}_{\text{sensor}}$$

Since rotation matrices are not easily interpreted, Euler angles are often used to describe rotation instead. These angles provide an intuitive and easy-to-visualize representation of orientation. However, for actual computations—such as constructing a transformation matrix—they are always converted back into a rotation matrix.

In practice, we:

1. Describe orientation using yaw, pitch, and roll
2. Convert those angles to a rotation matrix
3. Use the rotation matrix and translation vector to create the full transformation matrix
4. Apply the transformation to positions and directions

We conclude this section with a simple demonstration that illustrates how a pose can differ across various coordinate systems.

Listing 6.1 Use Euler angles to transform pose

```
import numpy as np
from scipy.spatial.transform import Rotation as R

# Define orientation in Euler angles
yaw, pitch, roll = 90, 0, 0 # in degrees

# Step 1: Convert to rotation matrix
rot_matrix = R.from_euler('zyx', [yaw, pitch, roll], degrees=True).as_matrix
    ()

# Step 2: Define translation vector
translation = np.array([5, 0, 0])

# Step 3: Build transformation matrix
T = np.eye(4)
T[:3, :3] = rot_matrix
T[:3, 3] = translation

# Step 4: Transform a position and a heading direction
position_sensor = np.array([1.0, 2.0, 0.0, 1.0])
heading_sensor = np.array([1.0, 0.0, 0.0]) # x-forward

position_world = T @ position_sensor
heading_world = rot_matrix @ heading_sensor

print("World␣position:", position_world[:3])
print("World␣heading:", heading_world)
```

6.3 Where GPS Falls Short

If you're thinking that if cars can just use GPS for localization, that's a fair question. After all, GPS is what we use every day to navigate with our phones. But for autonomous vehicles, it's not quite good enough.

Most consumer GPS systems are accurate to a few meters. That might be fine when you're walking to a coffee shop, but not when a car is trying to stay in its lane or stop exactly at a crosswalk. In cities, tall buildings can reflect signals and confuse the system. In tunnels or under dense trees, GPS signals might not come through at all.

That's why autonomous vehicles use a whole set of sensors—along with clever software—to localize themselves far more accurately than GPS alone ever could. Because GPS is prone to drift, blockage, and multipath interference, autonomous systems rely on a fusion of GPS, IMUs, LiDAR, and cameras to determine their exact location in real time.

To illustrate the limitations of raw GPS in WGS84 format, the following Python demo simulates a set of GPS points around a known location. Even slight position errors—common in real-world usage—can produce several meters of drift, making lane-level accuracy unreliable.

Dependencies: matplotlib, pyproj, numpy

Listing 6.2 Simulating and plotting GPS drift in WGS84

```
import matplotlib.pyplot as plt
from pyproj import Transformer
import numpy as np

# Sample GPS coordinates (latitude, longitude) simulating drift
gps_data = [
    (37.7749, -122.4194), # True position
    (37.77492, -122.41942),
    (37.77485, -122.41938),
    (37.77495, -122.41945),
    (37.77488, -122.41930)
]

# Convert WGS84 to UTM (Zone 10N for San Francisco)
transformer = Transformer.from_crs("epsg:4326", "epsg:32610", always_xy=True
    )
utm_coords = [transformer.transform(lon, lat) for lat, lon in gps_data]

# Extract and plot UTM coordinates
x, y = zip(*utm_coords)

plt.figure(figsize=(6, 6))
plt.plot(x, y, 'o-', label="GPS Drift (WGS84)")
plt.scatter(x[0], y[0], color='red', label='True Position')
plt.xlabel("Easting (m)")
plt.ylabel("Northing (m)")
plt.title("Simulated GPS Drift (WGS84)")
plt.legend()
plt.grid(True)
plt.axis("equal")
plt.show()
```

This code converts WGS84 lat/lon coordinates into UTM meters using pyproj, allowing us to visualize how GPS drift might appear in a real urban setting. Even within a few meters, the variation is enough to cross into adjacent lanes, which is unacceptable for safe autonomous navigation.

6.4 LiDAR-Based Localization

6.4.1 High-Precision Pose Tracking Using ICP

LiDAR's major advantage is its ability to produce high-precision, dense 3D point clouds of the environment. As illustrated in Fig. 6.1, a 3D point cloud captures surrounding houses, roadside trees, and road structures from an oblique view, while a bird's-eye view reveals the unique geometric layout of the road. These geometric features are key to performing LiDAR-based localization.

Currently, LiDAR-based pose tracking mainly relies on point cloud registration techniques. In essence, these techniques search for the best alignment between the map point cloud and the real-time LiDAR scan based on their geometric structures. The definition of "best alignment" can vary with different registration methods. For the classical Iterative Closest Point (ICP) algorithm [16], the best alignment is achieved when the sum of the distances from each point in the LiDAR scan to its nearest neighbor in the map point cloud is minimized. Figure 6.2 illustrates how the point cloud gradually moves toward the target position during the registration process.

As the LiDAR scan incrementally approaches the correct alignment, the algorithm stops when the matching error falls below a predefined threshold or after a fixed number of iterations. To help readers better understand this process, we

Fig. 6.1 Example of a LiDAR 3D point cloud from oblique and bird's-eye views

Fig. 6.2 Illustration of the ICP registration process, showing the iterative movement of the point cloud toward optimal alignment

provide an executable example, ICPDemo.py.[1] By running this program in the terminal, readers can replicate the iterative process shown in Fig. 6.2.

If one compares the initial positions of the two point clouds, they are already relatively close. Note that the demo includes several parameters to specify the initial translation and relative rotation. Next, we explore how the initial estimate affects the final localization result. First, keep the relative rotation fixed and vary only the initial translation using the following command:

```
python ICPDemo.py --init_translation [value] --init_rotation
[fixed_value]
```

As the initial translation increases, the number of iterations required by the algorithm grows. When the initial distance exceeds a certain threshold, the point cloud fails to converge; in fact, it gradually drifts away from the map region. Similarly, if we keep the translation constant and only alter the rotation with the command below:

```
python ICPDemo.py --init_translation [fixed_value] --init
_rotation [value]
```

In the case of rotation errors, the point cloud may not drift away, but it eventually converges to an incorrect position, unable to correct the rotational offset. These experiments highlight the importance of an accurate initial pose estimate in point cloud registration. While the initialization of the very first frame is addressed in the next section on global re-localization, here we assume that the initial pose is correctly known, and we discuss how to continuously update a reliable estimate. We consider two scenarios: high-speed and low-speed motions. Our dataset includes two sets of ten consecutive LiDAR scans along with their corresponding map coordinates for reference. Using ICPDemo.py, we test the localization of these ten frames when the initial estimate is the first frame's position. The reference commands are as follows:

```
python ICPDemo.py --dataset low_speed_dataset
python ICPDemo.py --dataset high_speed_dataset
```

For the low-speed scenario, the first four frames achieve a matching error of less than 0.1 meters. In contrast, for the high-speed scenario, only the first two frames meet this criterion. Clearly, whether at high or low speeds, the point cloud registration requires continuous updates of the initial estimate to ensure accurate matching. Two potential solutions include:

1. Using the previous frame's pose as the new initial estimate.
2. Predicting the next frame's pose by estimating the velocity from the first two frames.

As shown, both approaches successfully localize each frame with only minor differences in error relative to the reference coordinates. However, note that the

[1] The ICP demo: https://github.com/thecarlab/AVBook/blob/main/Chapter6/ICPDemo.py.

datasets used here were collected on a straight and flat road, where the inter-frame motion is predominantly translational. In scenarios involving significant rotations, such as during turns, failing to account for angular velocity can lead to notable deviations in the point cloud registration results.

In this section, we have demonstrated how to use the ICP algorithm for point cloud registration to align real-time LiDAR scans with a pre-built map. Through a series of experiments, we have shown that leveraging motion estimates to update the initial pose can significantly enhance the accuracy of point cloud registration under high-speed and turning conditions. It is worth noting that in the demo, the matching time for each frame often exceeds 2 seconds. For applications with strict real-time requirements, this performance is not acceptable. In the next section, we will introduce an alternative method—NDT (Normal Distributions Transform) registration—that achieves real-time localization.

6.4.2 Real-Time Pose Tracking Using NDT

In the previous section, we observed that the ICP algorithm, while effective, can be too slow for real-time pose tracking. In this demo, we introduce the Normal Distributions Transform (NDT) [12] method and demonstrate how applying a voxel grid filter to both the input LiDAR scan and the map can speed up computation. At the same time, we will discuss the trade-offs in estimation accuracy.

Before running the NDT registration, we first apply a voxel grid filter. This process divides the space into a grid of small cubes (voxels) and replaces all the points in each voxel with their centroid. This downsampling reduces the total number of points, making subsequent computations faster. However, keep in mind that filtering may also remove some of the fine details, which can slightly decrease the localization accuracy.

We provide a demo script named `NDTDemo.py`[2] that illustrates the entire process. To run the demo with a specified voxel size (e.g., 0.5 meters), use the following command in your terminal:

```
python NDTDemo.py --voxel_size 0.5
```

In this demo, the filtered LiDAR scan (displayed in blue) is matched against the filtered point cloud map (displayed in red). You will notice that the alignment happens much faster compared to processing the full, unfiltered data.

Experiment with different voxel sizes and observe the following:

- **Smaller Voxel Size:** More points are retained, which can lead to higher localization accuracy. However, the computational load increases, potentially slowing down the processing.

[2] The NDT demo: https://github.com/thecarlab/AVBook/blob/main/Chapter6/NDTDemo.py.

- **Larger Voxel Size:** Fewer points are used, which speeds up the registration process. On the downside, the reduced data density may lead to a slight loss in pose estimation precision.

This demo shows that NDT, combined with voxel grid filtering, offers a practical solution for achieving real-time pose tracking. It speeds up the registration process by reducing the number of points processed. However, this comes with a trade-off: while the computation is faster, the reduced detail in the data might slightly compromise accuracy. The key is to choose a voxel size that balances the need for real-time performance with acceptable localization precision for your specific application.

6.4.3 Global Re-localization with Scan Context and Map Keyframes

In real-world autonomous driving, the system may occasionally lose track of its precise location. To recover from such situations, global re-localization is essential. One effective approach for LiDAR-based re-localization leverages a unique point cloud descriptor called *Scan Context* along with the idea of *map keyframes*.

Scan Context is a method to summarize the structure of an entire LiDAR scan into a compact, image-like representation. Instead of comparing thousands of 3D points directly, Scan Context captures the spatial layout in a way that is robust to changes in viewpoint or rotation. This makes it much easier and faster to compare the current scan with previously stored scans.

Map keyframes are selected snapshots from the driving environment that represent significant or distinct locations. These keyframes are stored in a database along with their Scan Context descriptors and precise poses. When the system needs to re-localize, it simply extracts the Scan Context from the current scan and looks for a similar descriptor among the keyframes. The best match provides an initial estimate of the vehicle's global position.

While the core idea is simple, seeing it in action can be very enlightening. In our demo, you will observe that:

- The current LiDAR scan is converted into a Scan Context—a kind of "fingerprint" of the environment.
- This fingerprint is quickly compared against a database of fingerprints from map keyframes.
- The system identifies the most similar keyframe, which gives a good initial guess of the vehicle's location.

By running the demo, you will see how the Scan Context effectively narrows down the search space and helps the system to recover from localization failures, even in complex environments.

- *Efficiency:* The compact nature of the Scan Context descriptor allows for fast comparisons, making the re-localization process suitable for real-time applications.
- *Robustness:* Because it captures the overall layout of the surroundings, Scan Context remains effective even if some parts of the scene change or are temporarily occluded.
- *Dependence on Keyframes:* The accuracy of the global re-localization is tied to the quality and density of the map keyframes. More representative keyframes generally lead to better re-localization performance.

In summary, this approach helps the system to "reset" its position by matching the current environment against a library of known places. The demo you run will illustrate these concepts in action, showing how a well-designed Scan Context can bridge the gap between lost and recovered localization, and how map keyframes serve as reliable anchors in the environment.

6.5 Camera-Based Localization

In previous section, we have learned how to use LiDAR to localize the vehicle. Besides LiDAR, camera is also available for localization purpose. However, with directly measuring the distance to the environments, there is more work required to implement localization system based on camera. In this section, we will use a representative SLAM system as example, introducing how to obtain localization from camera.

ORB-SLAM utilizes visual cues from camera images, leveraging a combination of ORB feature extraction, map building, local tracking, and global re-localization:

- **ORB Feature Extraction:** Identifies unique visual points from images, creating robust feature descriptors.
- **Map Building:** Constructs and maintains a 3D representation of the environment through observed features.
- **Local Tracking:** Continuously estimates camera position relative to the recently built local map by matching observed features.
- **Global Re-localization:** Recognizes previously visited locations and recovers vehicle position after tracking loss by matching current features against the global map.

6.5.1 Image Feature Extraction and Matching

Visual SLAM systems rely on the ability to identify and track distinctive points in the environment across successive frames captured by a camera. These points, known as *image features*, provide the visual cues necessary to estimate the camera's motion and reconstruct the structure of the scene.

ORB-SLAM uses ORB (Oriented FAST and Rotated BRIEF) features for this purpose. ORB is a binary feature descriptor that combines the FAST corner detector with the BRIEF descriptor, enhanced with scale and rotation invariance. Compared to traditional descriptors such as SIFT or SURF, ORB is significantly faster and more computationally efficient, making it well-suited for real-time applications like autonomous driving.

The process begins with detecting keypoints in an image using the FAST detector. These are locations in the image where the local intensity sharply changes, typically at corners or edges. To ensure robustness to scale changes, ORB constructs an image pyramid, detecting keypoints at multiple resolutions. Each detected keypoint is then assigned a dominant orientation based on intensity gradients, allowing the system to achieve rotation invariance.

After detection, a binary descriptor is computed for each keypoint using BRIEF. This descriptor encodes the local appearance of the keypoint by performing intensity comparisons between pre-defined pairs of pixels within a patch centered at the keypoint. The resulting binary string is compact and efficient to match using the Hamming distance metric.

Once descriptors are extracted from two successive frames (or from a frame and a keyframe in the map), feature matching is performed. In ORB-SLAM, this is typically done using a brute-force matcher with Hamming distance and cross-check validation. Matches with the lowest distance are retained, and outliers are rejected through techniques like RANSAC. These correspondences form the basis for estimating the relative pose of the camera between frames.

The robustness of ORB features to changes in illumination, viewpoint, and partial occlusion makes them particularly effective in outdoor driving scenarios. However, they can still be affected by severe lighting conditions, motion blur, or dynamic objects. In such cases, parameter tuning (e.g., FAST threshold, number of features) or image preprocessing (e.g., histogram equalization) may be employed to enhance feature stability.

Demo: ORB Feature Extraction and Matching

To illustrate this process, we show a practical example using OpenCV's ORB implementation. Given two grayscale images from a forward-facing vehicle camera, the ORB detector is used to extract keypoints and descriptors, followed by brute-force matching.

Listing 6.3 ORB feature extraction and matching using OpenCV

```
import cv2
import matplotlib.pyplot as plt

# Load two grayscale images
img1 = cv2.imread('frame1.png', cv2.IMREAD_GRAYSCALE)
img2 = cv2.imread('frame2.png', cv2.IMREAD_GRAYSCALE)

# Initialize ORB detector
```

```
orb = cv2.ORB_create(nfeatures=1000)

# Detect and compute features
kp1, des1 = orb.detectAndCompute(img1, None)
kp2, des2 = orb.detectAndCompute(img2, None)

# Match features using Hamming distance
bf = cv2.BFMatcher(cv2.NORM_HAMMING, crossCheck=True)
matches = bf.match(des1, des2)
matches = sorted(matches, key=lambda x: x.distance)

# Draw top 50 matches
matched_img = cv2.drawMatches(img1, kp1, img2, kp2, matches[:50], None,
    flags=2)

# Display result
plt.imshow(matched_img)
plt.title("ORB␣Feature␣Matching")
plt.axis("off")
plt.show()
```

This example demonstrates how visual similarity between frames is captured through feature matching. These matches are later used to compute the relative pose between frames, which serves as the foundation for visual odometry in ORB-SLAM.

6.5.2 *Bag-of-Words and Loop Closing*

Visual localization methods rely on camera imagery to determine the vehicle's position within its environment. One widely used approach, ORB-SLAM, uses visual features to simultaneously build maps (SLAM) and estimate camera poses, effectively solving localization tasks. Beyond local tracking (odometry), ORB-SLAM also provides capabilities for global re-localization, critical for robustness in real-world applications.

The vehicle continuously tracks its current position by matching the ORB features detected in the live camera feed against a local subset of map points previously observed. This local tracking relies on frame-to-frame motion estimation (visual odometry), ensuring smooth and continuous localization.

Global re-localization is activated when local tracking fails—due to rapid motion, abrupt camera movements, or drastic scene changes—causing temporary localization loss. ORB-SLAM resolves this by performing global descriptor matching across the entire map database, quickly recognizing and recovering the camera's absolute position in the map, ensuring resilience to tracking interruptions.

Demonstration of ORB-SLAM Global Re-localization (Python)
Although complete ORB-SLAM involves sophisticated implementations, we can conceptually illustrate global re-localization through loop closure detection using Bag-of-Words (BoW) features:

Listing 6.4 Simplified global re-localization illustration

```
import cv2
import numpy as np

# Initialize ORB detector and descriptor
orb = cv2.ORB_create(1000)

# Suppose we have stored descriptors from keyframes
database_descriptors = [...] # pre-stored ORB descriptors from keyframes

# Load current frame for re-localization
current_frame = cv2.imread('current_frame.png', cv2.IMREAD_GRAYSCALE)
kp_curr, des_curr = orb.detectAndCompute(current_frame, None)

# Brute-force match current frame descriptors against the database
bf = cv2.BFMatcher(cv2.NORM_HAMMING, crossCheck=True)
best_score, best_frame_idx = 0, -1

for idx, des_ref in enumerate(database_descriptors):
    matches = bf.match(des_curr, des_ref)
    matches = sorted(matches, key=lambda x: x.distance)
    score = sum([m.distance for m in matches[:50]])

    if best_frame_idx == -1 or score < best_score:
        best_score = score
        best_frame_idx = idx

print(f"Matched frame index for global re-localization: {best_frame_idx}")
```

Note: The above example is a conceptual demonstration; a practical global re-localization implementation typically involves optimized visual vocabulary (Bag-of-Words) approaches to efficiently query large descriptor databases.

6.5.3 Limitations and Practical Solutions

ORB-SLAM, despite its strengths, encounters several significant limitations in practical environments:

1. **Sensitivity to Lighting Conditions:**
   ORB features degrade in low-light or highly dynamic lighting scenarios, reducing reliability.

Practical Solutions:

- **Sensor Fusion:** Integrate camera data with other sensors (LiDAR, IMU) to maintain localization robustness despite visual degradation.
- **Adaptive Exposure Control:** Dynamically adjust camera settings to maintain image quality across lighting variations.

2. **Difficulties in Texture-less Environments:**
   Lack of distinct visual features in environments with plain or repetitive surfaces reduces localization accuracy and stability.
   Practical Solutions:

- **Additional Sensors:** Employ depth sensors or LiDAR for geometric-based localization in feature-poor areas.
- **Active Lighting Techniques:** Introduce structured lighting to artificially generate detectable features.

3. **Motion Blur and Rapid Movements:**
   Rapid vehicle motions may introduce motion blur, causing feature tracking failure and localization drift.
   Practical Solutions:

- **High-Speed Cameras:** Utilize high frame-rate cameras to reduce motion blur effects.
- **IMU Integration:** Complement visual tracking with inertial measurements to stabilize pose estimation during rapid movements.

This subsection thoroughly explored ORB-SLAM for camera-based localization, emphasizing both online localization and global re-localization capabilities. We highlighted specific practical limitations, including sensitivity to lighting, texture-less environments, motion blur, and computational complexity, and discussed feasible strategies to address these challenges. Understanding these limitations and corresponding solutions ensures effective real-world application of ORB-SLAM in autonomous driving scenarios.

6.6 Sensor Fusion with the Kalman Filter

6.6.1 Why Fusion Matters in Localization

Each sensor has its strengths—and its weaknesses:

- **IMU** provides high-frequency motion data (acceleration, rotation) but drifts over time.
- **Camera** gives rich visual cues, but is sensitive to lighting and feature-poor environments.

- **LiDAR** offers accurate depth, but is expensive and may struggle with reflective or transparent surfaces.

By combining data from all three, we get the best of each:

- IMU for fast updates and short-term motion
- Camera for visual localization
- LiDAR for structure-aware positioning

This is where the **Kalman Filter** comes in.

6.6.2 What Is the Kalman Filter?

The Kalman Filter (KF) is a mathematical tool for **estimating the state** of a system over time by combining noisy measurements.

In localization, the "state" could be the vehicle's:

- Position (x, y)
- Velocity
- Orientation

At every step, the filter:

1. **Predicts** the next state using a motion model (e.g., based on IMU)
2. **Updates** the prediction using sensor measurements (LiDAR, camera)

It finds the best estimate by *balancing confidence* in the prediction and the measurements.

6.6.3 Simple Fusion Example in Python: LiDAR + Camera + IMU

Let's walk through a simple 2D fusion using a basic Kalman filter. We'll simulate noisy position measurements from LiDAR and camera, and motion from an IMU.

Goal Estimate the position of a vehicle moving along a straight path.

Listing 6.5 Simple Kalman filter fusion: LiDAR + Camera + IMU

```
import numpy as np
import matplotlib.pyplot as plt

np.random.seed(42)

# Simulation setup
timesteps = 50
```

```
true_pos = np.linspace(0, 10, timesteps)
imu_velocity = 0.2 # m/s, constant

# Generate noisy measurements
lidar_meas = true_pos + np.random.normal(0, 0.5, timesteps)
camera_meas = true_pos + np.random.normal(0, 0.3, timesteps)
imu_meas = np.full(timesteps, imu_velocity) + np.random.normal(0, 0.05,
     timesteps)

# Kalman Filter Setup
x = 0.0 # initial position
v = 0.0 # initial velocity
P = np.eye(2) # initial uncertainty
dt = 0.2 # time between steps

# Define matrices
F = np.array([[1, dt], [0, 1]]) # state transition
H = np.array([[1, 0]])        # measurement model (position only)
Q = np.array([[0.01, 0], [0, 0.1]]) # process noise
R_lidar = 0.25 # lidar noise variance
R_cam = 0.09  # camera\index{Camera} noise variance

estimates = []

for t in range(timesteps):
    # Predict step
    u = imu_meas[t]
    x_pred = F @ np.array([x, v]) + np.array([0.5 * dt**2 * u, dt * u])
    P = F @ P @ F.T + Q

    # Combine measurements (camera\index{Camera} and lidar) using simple
         average
    z = (lidar_meas[t] / R_lidar + camera_meas[t] / R_cam) / (1 / R_lidar +
         1 / R_cam)
    R_combined = 1 / (1 / R_lidar + 1 / R_cam)

    # Update step
    y = z - H @ x_pred
    S = H @ P @ H.T + R_combined
    K = P @ H.T @ np.linalg.inv(S)
    x_update = x_pred + K.flatten() * y
    P = (np.eye(2) - K @ H) @ P

    x, v = x_update
    estimates.append(x)

# Plotting
plt.figure(figsize=(10, 5))
plt.plot(true_pos, label="True␣Position", linewidth=2)
plt.plot(lidar_meas, 'o', label="LiDAR", alpha=0.5)
plt.plot(camera_meas, 's',␣label="Camera",␣alpha=0.5)
plt.plot(estimates,␣'-k',␣label="KF␣Estimate",␣linewidth=2)
plt.xlabel("Time␣Step")
plt.ylabel("Position␣(m)")
```

```
plt.title("Kalman Filter: Sensor Fusion")
plt.legend()
plt.grid(True)
plt.show()
```

What's Happening Here?

- The IMU provides predicted motion (velocity).
- LiDAR and camera provide noisy but direct position measurements.
- The Kalman filter fuses all three into a more accurate position estimate.

6.6.4 When to Use Kalman Filters (and Their Variants)

The classic Kalman filter assumes:

- Linear motion and measurement models
- Gaussian noise

In real autonomous driving systems, we often use:

- **Extended Kalman Filter (EKF)** for nonlinear models [2]
- **Unscented Kalman Filter (UKF)** for better nonlinearity handling
- **Factor graphs/Graph-SLAM** for more scalable and map-based estimation

But the fundamental principle is the same: **use all available sensor data to estimate the state as accurately as possible.** The Kalman filter provides a mathematically grounded way to combine noisy data from LiDAR, camera, and IMU. It lies at the core of many localization systems, helping to track the vehicle's pose in real time and smooth out uncertainty from individual sensors.

6.7 Recent Researches

In the previous sections, we explored foundational approaches to localization using various sensor modalities. As a critical component of autonomous driving, localization technology continues to evolve rapidly. Recent research has focused on three key areas: deep learning and end-to-end methods, robustness with new sensors, and real-time and scalable solutions. Below, we review the state-of-the-art in these domains, highlighting key techniques, advantages, and limitations.

6.7.1 Deep Learning and End-to-End Methods

Deep learning has transformed localization by enabling end-to-end systems that map raw sensor data directly to pose estimates, reducing reliance on handcrafted features. Notable approaches include DeepVO [17], which employs deep recurrent convolutional neural networks for monocular visual odometry, and MagicVO [8], which integrates CNNs with bidirectional LSTM for enhanced robustness. PoseNet [10] facilitates real-time 6-DOF camera relocalization, while GANVO [1] leverages generative adversarial networks for unsupervised learning. Despite these advancements, end-to-end models are often regarded as "black boxes" due to their limited interpretability, which makes diagnosing failures particularly challenging. Recent advancements, such as self-supervised deep visual odometry with online adaptation [11], further improve adaptability but highlight the need for better interpretability.

6.7.2 New Sensors For Challenging Conditions

LiDAR based and vision based localization in challenging conditions, such as low visibility or dynamic scenes, has driven the adoption of new sensors and fusion techniques. Thermal cameras, as explored in thermal-inertial odometry (TIO) [14], enhance performance in low-light or foggy environments. Event-based cameras [6], capturing brightness changes for low-latency data, are effective in high-speed scenarios [9, 4].

6.7.3 Real-Time and Scalability

Real-time performance and scalability are essential for deploying localization systems on resource-constrained platforms like autonomous vehicles and drones. Optimization-based Visual Inertial Odometry (VIO) with bounded-sized windows ensures constant processing times, making it suitable for real-time applications [13]. Other works, such as [5], focus on optimizing the structure of ORB-SLAM by leveraging an efficient Lucas-Kanade optical flow method and an enhanced Bag of Binary Words algorithm. Photo-SLAM [7] achieves real-time performance by combining explicit geometric features for efficient localization with a lightweight hyper primitives map and a Gaussian-Pyramid-based training method that progressively learns multi-level implicit photometric features, enabling fast and photorealistic mapping on resource-constrained embedded platforms like the Jetson AGX Orin.

6.8 Hands-on Exercises

6.8.1 Implementing a Global Relocalization System with LiDAR and Camera

Objective In this exercise, students will design and implement a global relocalization system capable of recovering a vehicle's position from a lost state using either LiDAR-based or camera-based place recognition methods. The goal is to compare the performance and robustness of these two approaches under various conditions.

Task Extend your localization pipeline to include a **global relocalization module** that activates when the system detects that the vehicle is "lost" (e.g., when pose uncertainty exceeds a predefined threshold or tracking is explicitly disrupted). Implement and integrate the following techniques:

- **LiDAR-based relocalization** using scancontext: Extract global descriptors from incoming LiDAR scans and match them against a database constructed from a reference trajectory to estimate position.
- **Camera-based relocalization** using bagofwords: Convert incoming images into BoW descriptors and perform image retrieval against a prebuilt visual database for place recognition.

Use simulated routes in CARLA with previously recorded maps and datasets for both LiDAR and camera modalities. Induce localization failures by manually displacing the vehicle or simulating sensor dropouts, then evaluate each method's ability to recover the correct pose.

Deliverable Submit your implementation along with a report that includes:

- Visualizations of successful and failed relocalization attempts for both LiDAR and camera methods.
- Quantitative comparisons: accuracy (e.g., Euclidean error from ground truth), success rate, and time to recovery.
- A brief discussion of the conditions where each method excels or fails (e.g., lighting changes, dynamic objects, viewpoint variation).

Be prepared to present your approach and results, and propose a strategy for combining LiDAR and camera cues for more robust global relocalization.

6.8.2 Sensor Fusion for Localization Accuracy

Objective In this exercise, students will use the CARLA simulator within the BlueICE framework to collect data from multiple sensors LiDAR, camera, and IMU—and evaluate how different combinations affect Kalman filter-based localization performance.

Task Starting from the base Kalman filter implementation in Listing 6.5, run controlled experiments using the same simulated trajectory but with varying sensor configurations (e.g., IMU only, LiDAR + IMU, Camera + IMU, LiDAR + Camera + IMU). For each setup, collect synchronized sensor data, feed it into your Kalman filter pipeline, and analyze the resulting localization accuracy by comparing the estimated trajectory to the ground truth.

Use the same simulation route and environmental settings to ensure fair comparison. Optionally, experiment with different noise models or sensor rates to observe their influence on filter performance.

Deliverable Submit a set of localization plots (ground-truth vs. estimated trajectories) for each sensor combination, along with a concise summary of your findings. Describe your tuning strategy, the filter behavior under each configuration, and which combination yielded the best localization accuracy. Be prepared to explain trade-offs in accuracy, robustness, and real-time feasibility.

6.8.3 Implementing an Extended Kalman Filter (EKF)

Objective This exercise builds on the Kalman filter script provided in Listing 6.5. Students will extend the implementation to support nonlinear system models using the Extended Kalman Filter (EKF) framework, enabling improved localization in more realistic vehicle motion scenarios.

Task Using Listing 6.5 as your starting point, implement an Extended Kalman Filter that can handle nonlinear vehicle dynamics and observation models. Replace or augment the linear prediction and update steps with versions that use Jacobian matrices to linearize around the current state estimate. Apply your EKF to sensor data collected from the CARLA simulator (e.g., IMU, GPS, and optionally LiDAR or camera observations).

Validate the EKF using a known trajectory in simulation. Compare the EKF performance against the original Kalman Filter, particularly under conditions with non-negligible system or measurement nonlinearity (e.g., high curvature turns, IMU drift, or delayed GPS updates).

Deliverable Submit your EKF implementation code along with plots comparing the EKF and standard KF estimated trajectories to the ground truth. Include a short written summary describing your nonlinear motion and observation models, how you computed the Jacobians, and key differences you observed in filter behavior or performance.

Recommended Papers to Read

1. Bresson, G., Alsayed, Z., Yu, L., & Glaser, S. (2017). Simultaneous localization and mapping: A survey of current trends in autonomous driving. IEEE Transactions on Intelligent Vehicles, 2(3), 194–220. [3]

2. Sellat, Q., & Ramasubramanian, K. (2022). Advanced techniques for perception and localization in autonomous driving systems: A survey. Optical Memory and Neural Networks, 31(2), 123–144. [15]
3. Zheng, S., Wang, J., Rizos, C., Ding, W., & El-Mowafy, A. (2023). Simultaneous localization and mapping (slam) for autonomous driving: Concept and analysis. Remote Sensing, 15(4), 1156. [18]

References

1. Almalıoğlu, Y., et al. (2019). GANVO: Unsupervised deep monocular visual odometry and depth estimation with generative adversarial networks. In *2019 International Conference on Robotics and Automation (ICRA), May 2019* (pp. 5474–5480). https://doi.org/10.1109/ICRA.2019.8793512
2. Barrau, A., & Bonnabel, S. (2015). An EKF-SLAM algorithm with consistency properties. Preprint. arXiv:1510.06263.
3. Bresson, G., et al. (2017). A survey of current trends in autonomous driving. *IEEE Transactions on Intelligent Vehicles*, *2*(3), 194–220.
4. Chamorro, W., Solà, J., & Andrade Cetto, J. (2022). Event-based line SLAM in real-time. *IEEE Robotics and Automation Letters*, *7*, 1–8. https://doi.org/10.1109/LRA.2022.3187266
5. Ferrera, M., et al. (2021). OV^{2}SLAM: A fully online and versatile visual SLAM for real-time applications. *IEEE Robotics and Automation Letters*, *6*, 1399.
6. Guan, W., et al. (2024). EVI-SAM: Robust, real-time, tightly-coupled event–visual–inertial state estimation and 3D dense mapping. *Advanced Intelligent Systems*, *6*, 2400243.
7. Huang, H., et al. (2024). Photo-SLAM: Real-time simultaneous localization and photorealistic mapping for monocular, stereo, and RGB-D cameras. In *Proceedings of the IEEE/CVF Conference on Computer Vision and Pattern Recognition*.
8. Jiao, J., et al. (2019). MagicVO: An end-to-end hybrid CNN and Bi-LSTM method for monocular visual odometry. *IEEE Access*, 1. https://doi.org/10.1109/ACCESS.2019.2926350
9. Jiao, J., et al. (2021). Comparing representations in tracking for event camera-based SLAM. In *Proceedings of the IEEE/CVF Conference on Computer Vision and Pattern Recognition, June 2021* (pp. 1369–1376). https://doi.org/10.1109/CVPRW53098.2021.00151
10. Li, M., et al. (2021). VNLSTM-PoseNet: A novel deep ConvNet for real-time 6-DOF camera relocalization in urban streets. *Geo-spatial Information Science*, *24*, 1–15. https://doi.org/10.1080/10095020.2021.1960779
11. Li, S., et al. (2020). Self-supervised deep visual odometry with online adaptation. In: *Proceedings of the IEEE/CVF Conference on Computer Vision and Pattern Recognition, June 2020* (pp. 6338–6347). https://doi.org/10.1109/CVPR42600.2020.00637
12. Magnusson, M. (2009, Dec.). *The three-dimensional normal-distributions transform — An efficient representation for registration, surface analysis, and loop detection*. PhD thesis.
13. Qin, T. Li, P., & Shen, S. (2017). VINS-Mono: A robust and versatile monocular visual- inertial state estimator. *IEEE Transactions on Robotics*. https://doi.org/10.1109/TRO.2018.2853729
14. Saputra, M. R. U., et al. DeepTIO: A deep thermal-inertial odometry with visual hallucination. *IEEE Robotics and Automation Letters*, *5*, 1. https://doi.org/10.1109/LRA.2020.2969170
15. Sellat, Q., & Ramasubramanian, K. (2022). Advanced techniques for perception and localization in autonomous driving systems: A survey. *Optical Memory and Neural Networks*, *31*(2), 123–144.
16. Vizzo, I., et al. (2023). KISS-ICP: In defense of point-to-point ICP – Simple, accurate, and robust registration if done the right way. *IEEE Robotics and Automation Letters (RA-L)*, *8*(2), 1029–1036. https://doi.org/10.1109/LRA.2023.3236571

17. Wang, S., et al. (2017). DeepVO: Towards end-to-end visual odometry with deep recurrent convolutional neural networks. In: *2017 IEEE International Conference on Robotics and Automation (ICRA), May 2017* (pp. 2043–2050). https://doi.org/10.1109/ICRA.2017.7989236
18. Zheng, S., et al. (2023). Simultaneous localization and mapping (SLAM) for autonomous driving: Concept and analysis. *Remote Sensing*, *15*(4), 1156.

Chapter 7
Path Planning and Decision-Making

7.1 Introduction

Autonomous driving requires robust planning and decision-making to navigate complex environments while ensuring safety, efficiency, and comfort. This section provides a comprehensive overview of planning methods, encompassing global path planning, local trajectory generation, behavioral decision-making frameworks, and associated strategic methodologies. These capabilities collectively allow autonomous vehicles to interpret their surroundings, evaluate potential maneuvers, and determine optimal driving actions in dynamic contexts.

Planning techniques for autonomous systems are typically categorized into **three** principal classes: search-based, sampling-based, and optimization-based methods. Each class is grounded in distinct algorithmic paradigms. Search-based methods rely on graph traversal techniques to explore feasible paths within discretized representations of the driving environment. Sampling-based approaches construct solutions by incrementally sampling the state space, favoring probabilistic completeness in high-dimensional domains. Optimization-based methods formulate planning as a constrained optimization problem, often leveraging numerical solvers to compute trajectories that balance multiple objectives such as safety margins, smoothness, and energy efficiency.

Through this taxonomy, this chapter lays the groundwork for understanding how autonomous systems synthesize global and local planning objectives with real-time environmental feedback. This includes how vehicles respond to static map constraints and dynamic elements such as other agents, traffic signals, and unpredictable obstacles. The planning and decision-making stack must operate in real time and under uncertainty, thus integrating closely with perception and control layers for continuous adaptation.

Contributors: Zhaofeng Tian

W. Shi, Y. He, *Introduction to Autonomous Driving*,
https://doi.org/10.1007/978-3-031-99485-2_7

This chapter will examine each planning paradigm in detail and explore how behavior planning frameworks interface with motion planning layers to yield coherent and context-aware driving behavior.

7.2 Global Path Planning

7.2.1 *Search-Based Global Path Planning*

Search-based planning methods operate by discretizing the environment into a graph representation, where nodes correspond to spatial locations and edges denote feasible transitions between them. These methods are especially effective for computing global paths from a defined start position to a goal, while respecting static obstacles and road geometry constraints (Fig. 7.1).

Dijkstra's algorithm [4] forms the foundational basis for search-based planning by exhaustively exploring paths using a priority queue to guarantee the shortest route. Despite its optimality, the algorithm's computational burden limits its practicality for large environments. A [12] enhances efficiency by integrating a heuristic function, commonly Euclidean or Manhattan distance, to guide exploration toward the goal more selectively.

Hybrid A [5, 6] extends A by incorporating continuous vehicle state representations and non-holonomic motion constraints, such as turning radius and steering limitations. This adaptation is particularly suited for autonomous driving, where vehicle kinematics significantly influence feasible maneuvering.

To apply search-based methods in practice, the environment is converted into a graph, wherein intersections and way points form nodes and drivable paths define edges. Cost functions are associated with each edge to account for dynamic feasibility, including lane shifts, speed restrictions, and obstacle avoidance. For large-scale environments, techniques like jump point search [11, 18] improve computational efficiency by reducing redundant node expansions and emphasizing critical way points.

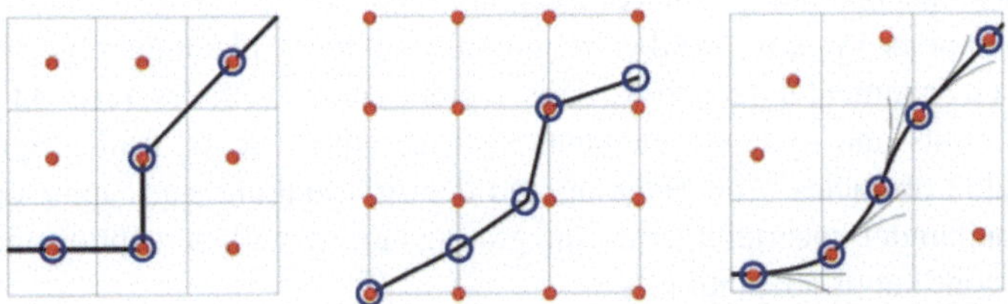

Fig. 7.1 Graphical comparison of search algorithms. Left: A* associates costs with centers of cells and only visits states that correspond to grid-cell centers. Center: Field D [7] associates costs with cell corners and allows any linear path from cell to cell. Right: Hybrid A associates a continuous state with each cell, and the score is the cost of its associated continuous state [5]

7.2.2 *Sampling-Based Global Path Planning*

Sampling-based methods, such as the Probabilistic Roadmap (PRM) [15] and Rapidly-exploring Random Tree (RRT) [16, 17], are widely utilized in high-dimensional and dynamically complex environments where explicit graph representations may be intractable. These methods are particularly adept at handling irregular obstacle distributions and discontinuous configuration spaces, offering a probabilistically complete approach that excels in scalability.

The PRM algorithm constructs a road map by randomly sampling collision-free configurations and connecting them to form a graph of feasible paths. Once constructed, standard search algorithms like A* are employed to query optimal paths between start and goal positions. PRM is most effective in static environments, allowing for road map reuse in repeated planning queries. Its key strength lies in its decoupled construction and query phases, enabling offline computation of the road map for rapid online response. However, its limitation emerges in dynamic or partially known environments where precomputed road maps may quickly become obsolete.

Conversely, RRT incrementally builds a search tree rooted at the initial state by iteratively sampling random points in the configuration space and extending the tree toward these samples, guided by feasibility constraints. RRT [14] enhances this approach by rewiring the tree to improve overall path cost, ensuring asymptotic optimality. Informed RRT [8] further narrows the search region using heuristic guidance, typically within an ellipsoidal domain, thereby improving computational efficiency. These adaptive variants are particularly useful when time constraints require rapid convergence toward feasible solutions, though they may produce non-smooth or jagged trajectories without post-processing.

To implement these methods, the vehicle initializes a representation of the environment, commonly using occupancy grids or point cloud-based maps. For PRM, nodes are sampled in free space and connected based on local visibility and dynamic feasibility. For RRT, the planner grows a tree structure incrementally, avoiding obstacles and adhering to kinematic constraints. In large-scale driving contexts, sampling efficiency is improved by employing the Frenet frame [28], which biases the sampling process along road-aligned directions, enhancing convergence and ensuring path feasibility. Despite these enhancements, the inherent randomness of sampling-based methods can lead to variability in performance, requiring multiple trials or integration with learning-based biasing strategies.

Sampling-based planners are probabilistically complete and offer scalability for complex environments, though post-processing is often required to smooth resulting paths and enforce dynamic consistency. Their flexibility and ease of implementation make them attractive for many scenarios, yet challenges remain in maintaining path optimality, repeatability, and integration with real-time constraints.

7.2.3 *Optimization-Based Path Planning*

Optimization-based methods explicitly incorporate vehicle dynamics and environmental constraints to compute feasible and efficient trajectories. These approaches formulate the path planning task as a constrained optimization problem, with objective functions that aim to minimize various performance metrics such as path length, travel time, energy consumption, curvature, or jerk, while satisfying hard constraints related to safety, road boundaries, and dynamic feasibility.

Gradient-based planners represent one major category within this class. These planners optimize cost functions defined by potential fields [22], Euclidean Signed Distance Fields (ESDF)[20, 29], or obstacle gradients[30], enabling smooth trajectory generation in cluttered environments. The main advantage of gradient-based methods is their ability to operate on continuous spaces, thereby avoiding discretization errors. However, they are often sensitive to local minima, particularly in environments with narrow passages or non-convex obstacles.

Convex optimization techniques address this issue by decomposing the environment into convex polytopes, allowing for global feasibility guarantees when constraints are convex [18]. These methods formulate the planning problem as a sequence of quadratic or linear programs, enabling fast computation and predictable performance. Yet, their applicability is restricted to scenarios where the environmental and vehicle constraints can be well-approximated in convex form.

Model Predictive Control (MPC)[1, 27] has gained significant attention due to its ability to plan and control simultaneously. MPC solves a receding-horizon optimization problem at each planning step, incorporating vehicle dynamics, control bounds, and collision avoidance within a single unified framework. Its real-time applicability makes it suitable for dynamic environments, although the computational load increases with problem complexity. Nonlinear MPC (NMPC)[9] extends this approach to handle vehicle nonlinearities, offering improved robustness and trajectory smoothness, albeit at higher computational cost and potential convergence issues.

Quadratic Programming (QP)-based approaches [23] also play a vital role in structured environments, such as highways and urban lanes. These planners define quadratic cost functions with linear or quadratic constraints to ensure smooth lane transitions, collision avoidance, and adherence to road boundaries. Their strength lies in producing optimal and smooth solutions under rigid kinematic constraints; however, QP formulations can struggle with dynamic changes in obstacle configurations unless frequently re-solved with updated inputs.

To implement optimization-based planning, the vehicle first extracts a local representation of the scene—typically a cost map, obstacle distance field, or Frenet-frame corridor. The planner then defines an objective function incorporating path quality metrics and feasibility constraints. Optimization solvers such as interior-point or active-set methods iteratively compute solutions, which are validated for physical consistency and safety.

Compared to search-based and sampling-based planners, optimization-based approaches offer a greater capacity for integrating vehicle dynamics, yielding smoother and more precise trajectories. Nevertheless, their real-time deployment depends critically on efficient solver performance, constraint convexity, and robust initialization. These limitations can be mitigated through warm-start strategies, hierarchical planning architectures, or hybrid methods combining global path guidance with local optimization refinement.

7.3 Local Trajectory Generation

Local trajectory generation plays a crucial role in ensuring safe, smooth, and efficient autonomous driving. The objective of local trajectory planning is to generate a short-term, dynamically feasible path that allows the vehicle to follow the global route while reacting to immediate obstacles, road conditions, and traffic regulations. Unlike global planning, which provides a high-level route from origin to destination, local planning must adapt in real-time to dynamic environments while ensuring passenger comfort and adherence to vehicle dynamics constraints.

A well-designed local trajectory generation framework must balance multiple factors, including smoothness, feasibility, safety, and computational efficiency. Smoothness is crucial for passenger comfort and vehicle longevity, requiring constraints on jerk and curvature. Feasibility ensures that the generated path respects kinematic and dynamic limits, such as acceleration and steering constraints. Safety considerations involve obstacle avoidance, maintaining safe distances from other vehicles, and adhering to traffic rules. Computational efficiency is essential for real-time execution, especially in high-speed scenarios where decisions must be made within milliseconds.

Various methodologies have been developed for local trajectory planning, each with different strengths and weaknesses. Frenet-based motion planning simplifies trajectory generation by using a curvilinear coordinate system aligned with the reference path, allowing efficient decoupling of lateral and longitudinal motion. Polynomial-based trajectory generation ensures smooth motion by using polynomial functions to interpolate between states while minimizing jerk. Optimization-based methods explicitly formulate trajectory generation as a constrained optimization problem, enabling precise control over multiple objectives.

The following subsections discuss these methods in detail, highlighting their mathematical foundations and practical applications in autonomous driving.

7.3.1 *Frenet-Based Motion Planning*

Frenet-based motion planning is widely used in autonomous driving, leveraging a curvilinear coordinate system aligned with the reference path [28]. This coordinate system represents vehicle motion in terms of longitudinal displacement s, which measures progress along the reference path, and lateral displacement d, which denotes deviation from the reference path.

Mathematically, a vehicle's position in Cartesian coordinates (x, y) can be mapped to the Frenet frame (s, d) using:

$$x = x_r(s) + d \cos(\theta_r(s) + \frac{\pi}{2}) \tag{7.1}$$

$$y = y_r(s) + d \sin(\theta_r(s) + \frac{\pi}{2}) \tag{7.2}$$

where $x_r(s)$, $y_r(s)$ define the reference path, and $\theta_r(s)$ is the orientation of the reference path at s. This formulation enables efficient motion planning by decoupling longitudinal speed control and lateral path deviation correction.

Frenet-based methods generate candidate trajectories by sampling different lateral offsets d at future time steps and optimizing them for smoothness, feasibility, and safety. These trajectories are evaluated using cost functions that penalize excessive jerk, abrupt lane changes, and proximity to obstacles. The optimal trajectory is selected based on minimizing the total cost while adhering to constraints such as acceleration and curvature limits [28]. As Fig. 7.2 shows candidates in different colors, with different cost functions, including the speed-oriented function in red, and the energy-efficient function in green [25].

This method is especially effective in structured environments like highways, where the road layout follows predefined lanes. Additionally, Frenet-based approaches allow easy integration with rule-based decision-making, such as lane-keeping and lane-changing maneuvers, enhancing overall motion planning efficiency.

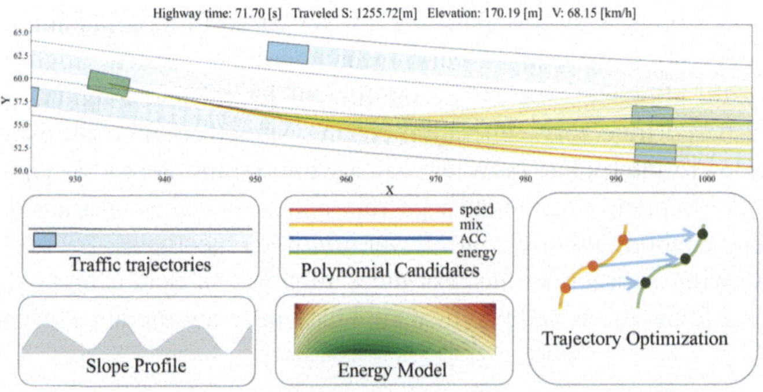

Fig. 7.2 Frenet polynomial trajectory candidates [25]

7.3.2 Polynomial-Based Trajectory Generation

Polynomial-based trajectory generation formulates motion as polynomial functions of time, ensuring smooth and continuous derivatives [28]. This method is commonly used in autonomous driving to generate feasible and comfortable trajectories by considering constraints on velocity, acceleration, jerk, and curvature.

A typical polynomial trajectory is represented as:

$$p(t) = a_0 + a_1 t + a_2 t^2 + a_3 t^3 + a_4 t^4 + a_5 t^5 \tag{7.3}$$

where coefficients $a_0, \ldots, a_5$ are determined based on initial and final conditions, such as position, velocity, and acceleration.

Quintic polynomials (fifth-order) are often used because they provide smooth transitions while satisfying boundary conditions. These trajectories minimize jerk, leading to comfortable motion and reducing wear on vehicle components. For lane changes and merging maneuvers, polynomial trajectories ensure seamless integration into traffic while maintaining safe distances from other vehicles.

Furthermore, polynomial-based planners can incorporate constraints such as maximum curvature κ_{max} and acceleration a_{max}, ensuring that generated paths comply with vehicle dynamics and safety regulations. Researchers have further enhanced this method by integrating polynomial optimization techniques, such as Bezier curves and B-splines, to improve robustness in highly dynamic environments [2, 3].

Compared to search-based and sampling-based methods, polynomial-based trajectory generation provides computational efficiency while maintaining smoothness and feasibility. However, it requires accurate boundary condition estimation, which can be challenging in highly dynamic environments where future states are uncertain.

Trajectory generation can be rigorously formulated as a constrained optimization problem, wherein the goal is to minimize a composite cost function subject to both vehicle dynamics and environmental constraints. This approach enables the planner to explicitly consider physical quantities such as jerk, acceleration, and curvature, thereby yielding trajectories that are not only dynamically feasible but also smooth and comfortable for passengers [30, 22, 20].

7.3.3 Balancing Feasibility, Smoothness, and Computational Cost

A well-constructed local trajectory must simultaneously satisfy three critical criteria: dynamic feasibility, smoothness, and computational tractability. Feasibility ensures compliance with the vehicle's kinematic and dynamic constraints, such as maximum acceleration, steering limits, and stability requirements. Smoothness, characterized by low values of jerk and curvature, contributes directly to ride comfort and vehicle durability. Computational efficiency is essential to ensure that the trajectory can be generated and updated in real time, especially in dynamic environments with fast-changing obstacle configurations.

To achieve this balance, trajectory planners often rely on a combination of numerical optimization techniques and heuristic pruning. Numerical methods allow for continuous-space optimization over complex cost landscapes, while heuristics reduce the dimensionality of the solution space or prioritize promising candidate trajectories, thereby enhancing real-time applicability.

7.3.4 Nonlinear Program for Trajectory Optimization

A general formulation of the trajectory optimization problem can be described as a nonlinear program (NLP), expressed in Eq. (7.4). This formulation includes a nonlinear cost function J and a set of constraints applied to both the vehicle states and control inputs. These constraints comprise both equality conditions—typically encoding the system's dynamics—and inequality conditions that ensure collision avoidance, limit violations, and terminal state requirements.

$$\begin{aligned}
\min_{(\mathbf{z}(t),\mathbf{u}(t))} \quad & J(\mathbf{z}(t), \mathbf{u}(t), [t_0, T]) \\
\text{s.t.} \quad & \dot{\mathbf{z}}(t) = f_z(\mathbf{z}(t), \mathbf{u}(t)), \\
& f_p(\mathbf{z}(t)) \not\subset \mathcal{W}_{\text{int}}, \forall t \in [t_0, T]; \\
& g_{\text{init}}(\mathbf{z}(t_0)) = 0, \\
& \underline{g}_t \leq g_{\text{itm}}(\mathbf{z}, \mathbf{u}) \leq \overline{g}_t, \\
& \underline{g}_T \leq g_{\text{end}}(\mathbf{z}(T)) \leq \overline{g}_T.
\end{aligned} \tag{7.4}$$

This mathematical formulation enforces that the state trajectory evolves according to the dynamic model, while remaining outside obstacle regions, satisfying initialization and terminal conditions, and adhering to additional path and control constraints.

A representative cost function integrating smoothness and energy efficiency is given in Eq. (7.5). This cost function penalizes deviation from desired velocity, excessive acceleration, high jerk, and close proximity to obstacles, while incorporating an energy term to reflect power consumption or control effort.

$$J = \int_{t_0}^{T} \Big(\underbrace{\|d_o(t), v(t) - v_d, a(t), j(t)\|_{2,\mathbf{w}_g}^2}_{\text{Smoothness Objectives}} + \underbrace{w_e \cdot f_e(t)}_{\text{Energy Efficiency}} \Big) dt, \tag{7.5}$$

where $\mathbf{w}_g = [w_o, w_v, w_a, w_j]$ denotes the weight vector applied to each component: obstacle distance d_o, velocity tracking error $v(t) - v_d$, acceleration $a(t)$, and jerk $j(t)$. Energy efficiency is weighted by w_e.

7.4 Decision Making in Autonomous Driving

Decision-making in autonomous vehicles involves the selection of appropriate actions based on sensor inputs, environmental constraints, and operational objectives. Robust decision-making is essential to ensure safety, efficiency, and adaptability in dynamic environments. Common decision-making paradigms in the autonomous driving domain include rule-based architectures, finite state machines (FSMs), reinforcement learning, and Markov Decision Processes (MDPs), each offering distinct advantages and trade-offs.

7.4.1 Markov Decision Processes (MDPs)

Markov Decision Processes (MDPs) provide a rigorous mathematical foundation for modeling sequential decision-making problems under uncertainty. An MDP is formally characterized by a tuple consisting of a state space, an action space, transition probability distributions, and a reward function. This structure enables the modeling of stochastic transitions and long-term utility of actions. In the context of autonomous driving, MDPs support the evaluation of expected cumulative rewards across action sequences, enabling optimization for safety, comfort, and energy efficiency. Partially Observable MDPs (POMDPs) extend this formulation by accounting for partial observability and sensor noise, thus enhancing robustness in real-world scenarios.

7.4.2 Decision Making in Autoware

Autoware, a widely used open-source autonomous driving software platform, implements a hierarchical decision-making framework. High-level behavior decisions such as stopping, accelerating, and lane-changing are derived based on fused sensory inputs and environmental understanding from the perception stack. Within Autoware, the behavior planning module utilizes FSMs coupled with cost-based evaluation strategies to select the most suitable maneuver. Trajectory planning modules ensure smooth transitions between decisions by incorporating motion constraints such as curvature limits and speed bounds. This layered approach facilitates reliable operation in diverse traffic conditions and supports modular integration with perception and control subsystems.

7.4.3 Behavior Planning and Decision Making

Behavior planning encompasses the selection of high-level driving maneuvers such as lane changes, merges, overtakes, and intersection stops. Effective behavior planning leverages a combination of structured logic and adaptive learning. Rule-based systems typically rely on FSMs and cost functions to evaluate and transition among predefined behavioral states. These systems offer transparency and deterministic behavior but may struggle with generalization in highly dynamic contexts.

Data-driven techniques, including deep reinforcement learning (DRL), enable autonomous agents to learn policies through interaction with either simulated or real-world environments [26]. These approaches have demonstrated adaptability in complex scenarios, although challenges related to sample efficiency and safety guarantees remain.

Hybrid strategies combine structured rule-based logic with probabilistic inference to improve decision robustness. POMDPs are particularly useful in modeling latent uncertainties in object tracking and agent intentions, thereby enhancing the reliability of decisions in occluded or ambiguous situations [19].

7.4.4 Narrow Passage Navigation

Navigating narrow and constrained environments such as corridors, tight intersections, or alleyways presents unique challenges due to limited clearance and high proximity to obstacles. As illustrated in Fig. 7.3, traditional planners may struggle in such settings when rigid obstacle avoidance constraints preclude feasible solutions.

The VOCAR framework addresses these limitations through a multi-layered strategy:

- Employing detailed environment mapping combined with topological path search algorithms to identify globally feasible navigation corridors.
- Dynamically adapting the robot's effective contour to account for tight spatial constraints, enabling more flexible local planning.

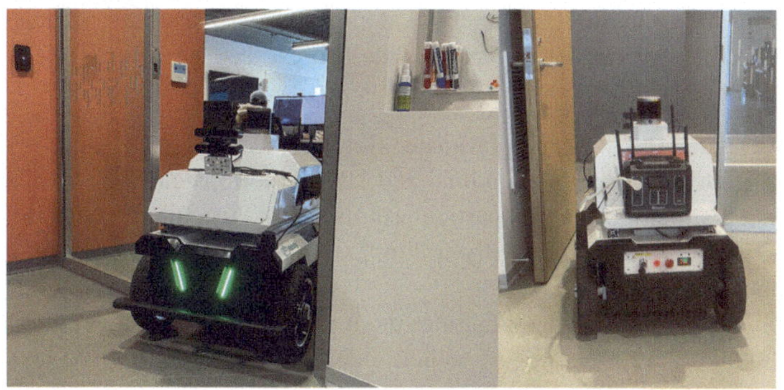

Fig. 7.3 A robot fails to pass an indoor narrow passage

- Integrating reinforcement learning-based local control policies trained to operate effectively in narrow environments.

7.5 Hands-on Exercises

7.5.1 Frenet Simulation

Trajectory planning is a fundamental component of autonomous driving, enabling vehicles to navigate safely and efficiently in dynamic environments. One of the widely used techniques for motion planning is the **Frenet-based approach**, which simplifies trajectory generation by representing the vehicle's motion in a curvilinear coordinate system. Instead of operating in traditional Cartesian coordinates, the Frenet frame describes motion in terms of **longitudinal progress** along a reference path and **lateral deviation** from it. The original paper can be referred to [25], and the whole code package can be downloaded at EMATO GitHub Repository.

This section presents a **practical implementation** of a Frenet-based trajectory planning algorithm. The provided Python function, `frenet_sim()`, simulates an ego vehicle navigating a structured road environment while interacting with surrounding traffic. The simulation includes key aspects such as:

- **Road Representation:** A predefined road with lane information and curvature is used to establish the reference path.
- **Traffic Modeling:** Multiple traffic vehicles with different speeds and positions are generated to simulate real-world driving scenarios.
- **Ego Vehicle Initialization:** The ego car starts from a given position and speed, and its future motion is planned based on a set of feasible trajectories.

- **Collision Awareness:** The function identifies relevant traffic vehicles in the vicinity of the ego car and determines the nearest leading vehicle to ensure safe following behavior.

By leveraging the Frenet coordinate system, the simulation allows **efficient trajectory generation**, as longitudinal and lateral motions can be optimized independently. This methodology is particularly useful in structured environments like highways, where vehicles primarily follow lane-based movements with controlled lane changes and speed adjustments.

The following code snippet illustrates the implementation of the `frenet_sim()` function, showcasing how an ego vehicle interacts with traffic while planning its future trajectory in the Frenet frame.

Listing 7.1 Frenet frame simulation

```
def frenet_sim(if_plot, car_type, g_type, r_type, if_use_frenet_policy):
    ws= 80
    v_list = []
    oft_v_list = []
    oft_s_d_list = []
    sim_time = 0
    total_time = 138
    time_step = 0.1
    recorder = Recorder()
    if_plot = if_plot
    if_use_frenet_policy = if_use_frenet_policy

    if car_type == 'truck':
        param = TruckParam()
    elif car_type == 'car':
        param = CarParam()
    param.gd_profile_type = g_type
    param.prediction_time = 5
    param.N = round(param.prediction_time/param.dt)+1
    param.safe_list = {'rsafe': 5, 'lon_safe': 5, 'lat_safe': 2.3}
    param.desired_v = 70/3.6
    param.desired_d = 0

    global_time = 0
    check_lb = -50
    check_ub = 100
    poft = None

    # Road initialization
    wx = [0.0, 350.0, 700.0, 1300.0, 1700.0, 2500,2900]
    wy = [0.0, 500.0, 150.0, 65.0, 0.0,-500,0]
    road = Road(wx, wy, num_lanes=3)

    profile_altitude, profile_gradient = get_gradient_profile(feature=param.
        gd_profile_type)

    # Simulated traffic cars
```

```
    tcars = []
    for bs in np.arange(0, 1800, 70):
        tcars.append(TrafficCar(lane=0, sd = [bs, 0], speed = 50/3.6, road =
             road))
    for bs in np.arange(30, 1800, 90):
        tcars.append(TrafficCar(lane=1, sd = [bs, 0], speed = 56/3.6, road =
             road))
    for bs in np.arange(40, 1800, 70):
        tcars.append(TrafficCar(lane=2, sd = [bs, 0], speed = 60/3.6, road =
             road))

    # Ego car initialization
    start_s = 100
    tcar_e = TrafficCar(lane=1, sd = [start_s,0],speed = 12.0, road = road)
    es, ed = tcar_e.s, tcar_e.d
    xc = [es, 70/3.6, 0, ed, 0, 0]

    # Traffic set up
    cared_index_list = []
    traffic_traj_sd_list = []
    leading_car = None
    leading_dist = np.inf

    for i in range(len(tcars)):
        if check_lb < tcars[i].s - es < check_ub:
            cared_index_list.append(i)
            tcars[i].calc_traffic_future_sd_points(param.prediction_time+
                 param.dt,param.dt)
            traffic_traj_sd_list.append(tcars[i].get_future_sd_points())
            tcars[i].set_xyyaw_with_sd()
            if abs(tcars[i].d - ed) < 0.6:
                if 5 < tcars[i].s - es < leading_dist:
                    leading_car = tcars[i]
                    leading_dist = tcars[i].s - es
```

After setting up the multi-lane simulation, the following code

Listing 7.2 Frenet trajectory generation

```
class FrenetTraj:
    def __init__(self,road, k, o, c):
        self.road = road
        # all trajectory, not number
        self.t = []
        self.d = []
        self.d_d = []
        self.d_dd = []
        self.d_ddd = []
        self.s = []
        self.s_d = []
        self.s_dd = []
        self.s_ddd = []
```

```
        self.clon= 0.0
        self.clat = 0.0
        self.cf = 0.0

        self.x = []
        self.y = []
        self.dx = []
        self.dy = []
        self.yaw = []
        # arc coordinate not lon
        self.dl = []
        self.l = []
        self.v = []
        self.a = []
        self.jerk = []
        self.k = []; self.k2 = []

        self.dfl = []
        self.fl = []
        self.fv = []
        self.fa = []
        self.fjerk = []
        self.fyaw = []
        self.fk = []; self.fk2 = []

        self.if_sd_collision = True
        self.if_xy_collision = True
        self.if_over_curvy = False
        self.if_over_jerky = False

        self.r_collision = np.inf
        self.r_curvature = np.inf
        self.r_jerk = np.inf
        self.r_invalid = None
        self.r_v = None
        self.r_d = None
        self.r_fe = None
        self.r_complex1 = None
        self.r_complex2 = None
        self.r_complex3 = None

        self.k,self.o,self.c = k,o,c

    def setup(self,xc,xg,road,ptime,dt, gd_profile):
                # (xs, vxs, axs, xe, vxe, axe, time,dt):
            cs,cs_d,cs_dd,cd,cd_d,cd_dd = xc
            gs,gs_d,gs_dd,gd, gd_d, gd_dd = xg
            Tlat = QuinticPolynomial(cd, cd_d, cd_dd, gd, gd_d,gd_dd, ptime,
                 dt)
            Tlon = QuinticPolynomial(cs,cs_d,cs_dd, gs, gs_d,gs_dd, ptime, dt
                 )
            self.t, self.d, self.d_d, self.d_dd, self.d_ddd = Tlat.get_traj()
            _, self.s, self.s_d, self.s_dd, self.s_ddd = Tlon.get_traj()
```

```
# assert len(self.s) == 50, " len is not 50!! len is " + str(len(
    self.s))

# ************* Based on x, y
self.x, self.y, _ = road.frenet_to_global(self.s, self.d)
self.dx = np.diff(self.x); self.dx = np.append(self.dx, self.dx
    [-1])
self.dy = np.diff(self.y); self.dy = np.append(self.dy, self.dy
    [-1])
self.yaw = np.arctan2(self.dy, self.dx)
self.dl = np.hypot(self.dx,self.dy)
self.l = np.cumsum(self.dl); self.l = np.append(0, self.l[:-1])
# Calculate velocity along the arc
self.v = self.dl / dt
# Calculate acceleration along the arc
self.a = np.diff(self.v) / dt
self.a = np.append(self.a, self.a[-1]) # Extend to match length
# Calculate jerk along the arc
self.jerk = np.diff(self.a) / dt
self.jerk = np.append(self.jerk, self.jerk[-1]) # Extend to match
     length
ddx = np.diff(self.dx); ddx = np.append(ddx, ddx[-1])
ddy = np.diff(self.dy); ddy = np.append(ddy, ddy[-1])
diff_yaw = np.diff(self.yaw); diff_yaw = np.append(diff_yaw,
    diff_yaw[-1])
self.cur = diff_yaw/self.dl
self.cur2 = np.abs(self.dx * ddy - self.dy * ddx) / (self.dx**2 +
     self.dy**2)**(3/2)

# *********** Based on s, d
self.dfl = np.hypot(self.d_d*dt, self.s_d*dt)
self.fl = np.cumsum(self.dfl)
self.fv = np.hypot(self.s_d, self.d_d)
self.fa = np.hypot(self.s_dd, self.d_dd)
self.fjerk = np.hypot(self.s_ddd, self.d_ddd)
self.fyaw = np.arctan2(self.s_d*dt, self.d_d*dt)

diff_fyaw = np.diff(self.fyaw); diff_yaw = np.append(diff_fyaw,
    diff_fyaw[-1])
self.fcur = diff_yaw/self.dfl
self.fcur2 = np.abs(self.s_dd * self.d_d - self.s_d * self.d_dd)
    / (self.s_d**2 + self.d_d**2)**(3/2)

self.theta = get_traj_theta(gd_profile, self.s)
self.sin = np.sin(self.theta)
self.cos = np.cos(self.theta)

self.ar = self.k[0]*self.v**2 + self.k[1]*self.cos + self.k[2]*
    self.sin
atemp = self.a + self.ar
self.at = np.where(atemp<0,0,atemp)
self.ab = self.at-self.a-self.ar
```

```
        self.fr = self.o[0] + self.o[1]*self.v + self.o[2]*self.v**2+self
            .o[3]* self.v**3 + self.o[4]*self.v**2+(self.c[0]+self.c[1]*
            self.v + self.c[2]*self.v**2)*self.at
        self.fc = np.cumsum(self.fr*dt) # this is total consumption of
            the traj
        self.fe_ds = self.fr *dt / self.s_d * dt
        self.fe = self.fr*dt / self.dl

        self.far = self.k[0]*self.fv**2 + self.k[1]*self.cos + self.k[2]*
            self.sin
        atemp = self.fa + self.far
        self.fat = np.where(atemp<0,0,atemp)
        self.fab = self.fat-self.fa-self.far
        self.ffr = self.o[0] + self.o[1]*self.fv + self.o[2]*self.fv**2+
            self.o[3]* self.fv**3 + self.o[4]*self.fv**2+(self.c[0]+self
            .c[1]*self.fv + self.c[2]*self.fv**2)*self.fat
        self.ffc = np.cumsum(self.ffr*dt)
        self.ffe_ds = self.ffr*dt / self.s_d * dt
        self.ffe = self.ffr*dt / self.dfl # ml/m to L/100 km

def calc_reward(self, traj_traffic_list, safe_list, jerkmax, kmax,
     desired_v, desired_d):
    """
    This is function only calculates the basic reward component for
        frenet algorithm,
    for more complex objective settings, alg.frenet.py will handle them.
    """

    """
    Collision
    """
    self.if_sd_collision, self.if_xy_collision= self.check_collision(
        traj_traffic_list, safe_list)
    if self.if_xy_collision:
        self.r_collision = np.inf
    else:
        self.r_collision = 0

    """
    Curvature
    """
    if any(abs(self.cur) > 0.1) :
        self.if_over_curvy = True
        self.r_curvature = np.inf
    else:
        self.if_over_curvy = False
        self.r_curvature = 0

    """
    Jerk
    """
    self.flon_jerk = self.s_ddd**2
    self.flat_jerk = self.d_ddd**2
    # print("flon_jerk: ", self.flon_jerk)
```

```
        # print("flat_jerk: ", self.flat_jerk)
        # print(np.max(self.flon_jerk + self.flat_jerk) )
        if np.max(self.flon_jerk + self.flat_jerk) > jerkmax**2 + jerkmax**2:
            self.if_over_jerky = True
            self.r_jerk = np.inf
        else:
            self.if_over_jerky = False
            self.r_jerk = np.sum(self.flon_jerk + self.flat_jerk)

        """
        End state: Lon s_d (longitute v) and Lat d
        """
        self.r_v = (self.s_d[-1] - desired_v)**2

        self.r_v = np.sum((self.s_d -desired_v)**2)
        self.r_d = (self.d[-1] - desired_d)**2

        """
        Fuel efficiency:
        fuel consumption of the trajectory / traveled s (not arc length but
             solely length on s)
        """
        self.r_fe = self.fc[-1] / (self.s[-1] - self.s[0])

        self.r_invalid = self.r_collision+ self.r_curvature +self.r_jerk

        # self.r_complex1 = 0.001* self.r_jerk + self.r_v + 0.1*self.r_d
        self.r_complex1 = 0.00* self.r_jerk + self.r_v
        self.r_complex2 = 0.001*self.r_jerk + self.r_v + 100* self.r_fe

        # assert len(self.r_fe) == 1, "rfe len"
        # assert len(self.r_complex2) == 1, "length wrong"

        self.r_complex3 = self.r_fe

    def check_collision(self,traj_traffic_list, safe_list):
        if len(traj_traffic_list) == 0:
            return(False, False)

        else:

            sd_count = 0
            xy_count = 0
            for traj in traj_traffic_list:
                traffic_s, traffic_d = traj[:,0],traj[:,1]
                traffic_x, traffic_y, _ = self.road.frenet_to_global(traffic_s
                     , traffic_d)
                sd_dist_square = (traffic_s - self.s)**2 + (traffic_d - self.d
                     )**2
                xy_dist_square = (traffic_x - self.x)**2 + (traffic_y - self.y
                     )**2
                if np.min(np.abs(traffic_s -self.s)) < safe_list['lon_safe']
                     and np.min(np.abs(traffic_d - self.d)) < safe_list['
                     lat_safe']:
```

```
            sd_count += 1
            # print("min ds {} dd {}".format(np.min(np.abs(traffic_s -
                self.s)), np.min(np.abs(traffic_d - self.d))))
            # print("min dd ",np.min(np.abs(traffic_d -self.d)))
        # print("sd_dist_squre: ", sd_dist_square)
        # print("xy_dist_squre: ", xy_dist_square)

        # if np.min(sd_dist_square) < safe_list["rsafe"]**2:
        #     sd_count += 1
        #     # print("sd collsion!!!!")

        if np.min(xy_dist_square) < safe_list['rsafe']**2:
            xy_count += 1

    return(sd_count>0, xy_count>0)
```

Listing 7.3 Nonlinear trajectory optimization

```
"""
Boundary Value Problem NLP
"""
class EMATO:
    def __init__(self, param):
        self.id = "NLP"
        self.nx = 2 ; self.nu = 1
        self.vr = param.vr ; self.xr = param.xr
        self.xmin = param.xmin.copy() ; self.xmax = param.xmax.copy()
        self.umin = param.umin ; self.umax = param.umax
        self.vmin = param.vmin; self.vmax = param.vmax
        self.bmin = param.bmin; self.bmax = param.bmax
        self.w1 = param.w1 ;self.w2 = param.w2 ;self.w3 = param.w3
        self.T = param.T ; self.dmin = param.dmin; self.dmax = param.dmax
        self.N = param.N ; self.dt = param.dt ; self.dinit = param.dinit
        self.param = param
        self.theta = 0.
        #  self.gd_profile = None
        _, self.gd_profile = get_gradient_profile(feature=param.
            gd_profile_type)
        self.use_gd_prediction = param.use_gd_prediction

        self.k1,self.k2,self.k3 = param.k
        # self.b0, self.b1, self.b2, self.b3 = param.b
        self.c0,self.c1,self.c2 = param.c
        self.o0,self.o1,self.o2,self.o3,self.o4 = param.o

        self.xs = None
        self.xe = None
        print("xs␣,␣xe:␣", self.xs, self.xe)

        self.XS = ca.MX.sym('XS', 3)
        self.XE = ca.MX.sym('XE', 3)

        # self.sec = ca.MX.sym('sec', 1)
```

```
# self.vec = ca.MX.sym('vec', 1)
if self.use_gd_prediction:
    self.gdec = ca.MX.sym('gdec', self.N)
else:
    self.gdec = ca.MX.sym('gdec', 1)

self.p = ca.vertcat(self.XS, self.XE,self.gdec)
self.S = ca.MX.sym('S', self.N)
self.U = ca.MX.sym('U', self.N) # Control input vector
self.V = ca.MX.sym('V', self.N) # Velocity vector
self.A = ca.MX.sym('A', self.N) # Apparent Acceleration
self.B = ca.MX.sym('B', self.N)
self.Jerk = ca.MX.sym('Jerk', self.N)

self.p_value = None

self.fu = self.o0 + self.o1 * self.V + self.o2 * self.V**2 + \
            self.o3 * self.V**3 +self. o4 * self.V**4 + \
                (self.c0 + self.c1 * self.V + self.c2 * self.V**2) *
                    self.U

self.J = self.w1 * ca.sum1(self.Jerk**2) +\
        self.w2 * ca.sum1(self.A**2 + self.B**2)+ \
        self.w3 *ca.sum1(self.fu)/ (self.S[-1] - self.S[0])
"""
Constraints
"""

# Dynamic Constraints: [N]
self.gfa = self.A- ( self.U - self.B- self.k1 * self.V**2 - self.k2 *
     ca.cos(self.gdec) - self.k3 * ca.sin(self.gdec) )
self.gfv = [] ; self.gfd = []; self.gfj = []

# V dynamic: [N-1]
for i in range(self.N-1):
    self.gfv.append(self.V[i+1] - (self.V[i] + self.A[i] * self.dt))
# A dynamic: [N-1]
for i in range(self.N-1):
    self.gfd.append(self.S[i+1]-(self.S[i] + self.V[i]*self.dt + 0.5*
         self.A[i]* self.dt**2))
# Jerk dynamic: [N-1]
for i in range(self.N-1):
    self.gfj.append(self.A[i+1]-(self.A[i]+self.Jerk[i]*self.dt))

# Boundary Value: [6]

self.gss = self.S[0] - self.XS[0]
self.gse = self.S[-1] - self.XE[0]
self.gvs = self.V[0] - self.XS[1]
self.gas = self.A[0] - self.XS[2]
self.gve = self.V[-1] - self.XE[1]
self.gae = self.A[-1] - self.XE[2]

"""
```

```
Concat all the defined constraints
"""
self.g = ca.vertcat(self.gfa,*self.gfv,*self.gfd, *self.gfj, self.gss
    , self.gvs, self.gas, self.gse, self.gve, self.gae)
# self.g = ca.vertcat(self.gacc, self.gfa,*self.gfv,*self.gfd, *self.
    gfj, self.gss,self.gse)
# self.g = ca.vertcat(self.gfa,*self.gfv,*self.gfd, *self.gfj, self.
    gss,self.gse)
# self.g = ca.vertcat(self.gacc, self.gfa,*self.gfv,*self.gfd, *self.
    gfj)
self.nlp = {'x': ca.vertcat(self.S, self.V,self.U,self.Jerk, self.A,
    self.B), 'f': self.J, 'g': self.g, 'p':self.p}
self.opts = {
    'ipopt.print_level': 0, # Adjust verbosity (0 to 12, with 0 being
        silent and 12 being very verbose)
    'print_time': True, # Print solver execution time
    'ipopt.tol': 1e-8, # Tolerance (adjust as needed)
}
self.solver = ca.nlpsol('solver', 'ipopt', self.nlp, self.opts)
# Initial guess and bounds

# Lower bounds for v and u
self.lb_s = -np.inf * np.ones(self.N)
self.lb_v = np.zeros(self.N) # v >= 0
self.lb_u = self.umin * np.ones(self.N) # u >= -1
self.lb_jerk = - self.param.jerkmax * np.ones(self.N)
self.lb_a = -3.0 * np.ones(self.N)
self.lb_b = self.bmin * np.ones(self.N)
self.lbx = np.concatenate((self.lb_s, self.lb_v, self.lb_u, self.
    lb_jerk, self.lb_a, self.lb_b))

self.ub_s = np.inf * np.ones(self.N)
self.ub_v = 27.* np.ones(self.N) # v >= 0
self.ub_u = self.umax * np.ones(self.N) # u >= -1
self.ub_jerk = self.param.jerkmax * np.ones(self.N)
self.ub_a = 2.0 * np.ones(self.N)
self.ub_b = self.bmax * np.ones(self.N)
self.ubx = np.concatenate((self.ub_s, self.ub_v, self.ub_u, self.
    ub_jerk,self.ub_a, self.ub_b))

# self.lbg = np.zeros(4*self.N+3) # Lower bounds of g
# self.ubg = np.concatenate(( self.dmax * np.ones(self.N), np.zeros
    (4*self.N+3))) # Upper bounds of g

self.lbg = np.zeros(4*self.N+3) # Lower bounds of g
self.ubg = np.zeros(4*self.N+3) # Upper bounds of g

# self.lbg = np.zeros(5*self.N-3) # Lower bounds of g
# self.ubg = np.concatenate(( self.dmax * np.ones(self.N), np.zeros
    (4*self.N-3))) # Upper bounds of g

assert len(self.lbg) == len(self.ubg), "wrong length for box 
    constraints"
self.solve_time = 0.
```

```
    # s, v, u, jerk, a, b where u:= at, a:= av
    def solve(self):
        t1 = time.time()
        # assert len(self.init_guess) == 300 , " Guess wrong len: "+str(len(
            self.init_guess))
        res = self.solver(x0=self.init_guess, lbx=self.lbx, ubx=self.ubx,\
                    lbg = self.lbg, ubg = self.ubg, p = self.p_value)
        self.res = np.array(res['x']).flatten()
        t2 = time.time()
        self.solve_time = t2-t1

        traj_s = self.res[:self.N]
        traj_v = self.res[self.N: 2*self.N]
        # print("Traj_s: ", traj_s)
        # print("if s increasing: ", all(np.diff(traj_s)>= 0) )
        # print("Traj_v: ", traj_v)
        assert all(np.diff(traj_s)>= -0.01) , "␣going␣backward␣detected!!"
        assert all(traj_v) >= -0.1, "␣V␣should␣bigger␣than␣0!!"

        at = self.res[2*self.N]
        jerk = self.res[3*self.N]
        av = self.res[4*self.N]
        ab = self.res[5*self.N]
        # print("at: ", at, " av: ", av , "ab: ", ab)
        return self.get_res()

    def update(self, ft, if_dynamic_vr = True):
        self.xs = np.array([ft.l[0], ft.v[0], ft.a[0]])
        self.xe = np.array([ft.l[-1], ft.v[-1], ft.a[-1]])

        self.init_guess = np.concatenate((ft.l,\
                                ft.v,\
                                ft.at,\
                                ft.jerk, \
                                ft.a,\
                                ft.ab ))
        if self.use_gd_prediction:
            # s_index = np.concatenate((resx[1:self.N], [resx[self.N-1]]))
            s_index = ft.s.copy()
            index_gd = np.round(s_index).astype(np.int)
            assert len(index_gd) == self.N
            p_gd = self.gd_profile[index_gd]
            self.p_value = ca.vertcat(self.xs, self.xe, p_gd)
        else:

            self.p_value = ca.vertcat(self.xs, self.xe, 0)

    def get_res(self):
        # return self.res['x']
        return self.res

    def get_traj_s(self):
        return self.res[:self.N]
```

```
    def get_solve_info(self):
        return {'solve_time': self.solve_time, 'executed_its': 1}

    def set_gd_profile(self,gd_profile):
        self.gd_profile = gd_profile
```

Recommended Papers to Read

1. Vu, T. M., Moezzi, R., Cyrus, J., & Hlava, J. (2021). Model predictive control for autonomous driving vehicles. Electronics, 10(21), 2593. [27]
2. Huang, Y., Du, J., Yang, Z., Zhou, Z., Zhang, L., & Chen, H. (2022). A survey on trajectory-prediction methods for autonomous driving. IEEE Transactions on Intelligent Vehicles, 7(3), 652–674. [13]
3. Teng, S., Hu, X., Deng, P., Li, B., Li, Y., Ai, Y., ... & Chen, L. (2023). Motion planning for autonomous driving: The state of the art and future perspectives. IEEE Transactions on Intelligent Vehicles, 8(6), 3692–3711. [24]
4. Guo, Y., Guo, Z., Wang, Y., Yao, D., Li, B., & Li, L. (2023). A survey of trajectory planning methods for autonomous driving—Part I: Unstructured scenarios. IEEE Transactions on Intelligent Vehicles. [10]
5. Reda, M., Onsy, A., Haikal, A. Y., & Ghanbari, A. (2024). Path planning algorithms in the autonomous driving system: A comprehensive review. Robotics and Autonomous Systems, 174, 104630. [21]

References

1. Abbas, M. A., Milman, R., & Eklund, J. M. (2017). Obstacle avoidance in real time with nonlinear model predictive control of autonomous vehicles. *Canadian Journal of Electrical and Computer Engineering*, *40*(1), 12–22.
2. Chen, J., et al. (2013). Lane change path planning based on piecewise bezier curve for autonomous vehicle. In *Proceedings of 2013 IEEE International Conference on Vehicular Electronics and Safety* (pp. 17–22). IEEE.
3. Choi, J., et al. (2024). Safe and efficient trajectory optimization for autonomous vehicles using b-spline with incremental path flattening. *IEEE Transactions on Intelligent Transportation Systems*, *26*, 1797–1811.
4. Dijkstra, E. W. (2022). A note on two problems in connexion with graphs. In *Edsger Wybe Dijkstra: His Life, Work, and Legacy* (pp. 287–290).
5. Dolgov, D., et al. (2008). Practical search techniques in path planning for autonomous driving. *Ann Arbor*, *1001*(48105), 18–80.
6. Dolgov, D., et al. (2010). Path planning for autonomous vehicles in unknown semi-structured environments. *The International Journal of Robotics Research*, *29*(5), 485–501.
7. Ferguson, D., & Stentz, A. (2007). Field D*: An interpolation-based path planner and replanner. In *Robotics Research: Results of the 12th International Symposium ISRR* (pp. 239–253). Springer.
8. Gammell, J. D., Srinivasa, S. S., & Barfoot, T. D. (2014). Informed RRT*: Optimal sampling-based path planning focused via direct sampling of an admissible ellipsoidal heuristic. In *2014 IEEE/RSJ International Conference on Intelligent Robots and Systems* (pp. 2997–3004). IEEE.
9. Grüne, L., et al. (2017). *Nonlinear model predictive control*. Springer.

10. Guo, Y., et al. (2023). A survey of trajectory planning methods for autonomous driving—Part I: Unstructured scenarios. *IEEE Transactions on Intelligent Vehicles, 9*, 5407–5434.
11. Harabor, D., & Grastien, A. (2011). Online graph pruning for pathfinding on grid maps. In *Proceedings of the AAAI Conference on Artificial Intelligence* (Vol. 25, pp. 1114–1119).
12. Hart, P. E., Nilsson, N. J., & Raphael, B. (1968). A formal basis for the heuristic determination of minimum cost paths. *IEEE Transactions on Systems Science and Cybernetics, 4*(2), 100–107.
13. Huang,Y., et al. (2022). A survey on trajectory-prediction methods for autonomous driving. *IEEE Transactions on Intelligent Vehicles, 7*(3), 652–674.
14. Karaman, S., & Frazzoli, E. (2011). Sampling-based algorithms for optimal motion planning. *The International Journal of Robotics Research, 30*(7), 846–894.
15. Kavraki, L. E., et al. (1996). Probabilistic roadmaps for path planning in high-dimensional configuration spaces. *IEEE Transactions on Robotics and Automation, 12*(4), 566–580.
16. LaValle, S. (1998). *Rapidly-exploring random trees: A new tool for path planning*. Research Report 9811.
17. LaValle, S. M., & Kuffner, J. J. (2001). Rapidly-exploring random trees: Progress and prospects. In *Algorithmic and Computational Robotics* (pp. 303–307).
18. Liu, S., et al. (2017). Planning dynamically feasible trajectories for quadrotors using safe flight corridors in 3-d complex environments. *IEEE Robotics and Automation Letters, 2*(3), 1688–1695.
19. Lovejoy, W. S. (1991). A survey of algorithmic methods for partially observed Markov decision processes. *Annals of Operations Research, 28*(1), 47–65.
20. Ratliff, N., et al. (2009). CHOMP: Gradient optimization techniques for efficient motion planning. In *2009 IEEE International Conference on Robotics and Automation* (pp. 489–494).
21. Reda, M., et al. (2024). Path planning algorithms in the autonomous driving system: A comprehensive review. *Robotics and Autonomous Systems, 174*, 104630.
22. Rösmann, C., Hoffmann, F., & Bertram, T. (2017). Integrated online trajectory planning and optimization in distinctive topologies. *Robotics and Autonomous Systems, 88*, 142–153.
23. Stellato, B., et al. (2020). OSQP: An operator splitting solver for quadratic programs. *Mathematical Programming Computation, 12*(4), 637–672.
24. Teng, S., et al. (2023). Motion planning for autonomous driving: The state of the art and future perspectives. *IEEE Transactions on Intelligent Vehicles, 8*(6), 3692–3711.
25. Tian, Z., Xia, L., & Shi, W. (2024). *EMATO: Energy-model-aware trajectory optimization for autonomous driving*. Preprint. arXiv:2412.08830.
26. Tian, Z., et al. (2024). Unguided self-exploration in narrow spaces with safety region enhanced reinforcement learning for Ackermann-steering robots. In *2024 IEEE International Conference on Mobility, Operations, Services and Technologies (MOST)* (pp. 260–268). IEEE.
27. Vu, T. M., et al. (2021). Model predictive control for autonomous driving vehicles. *Electronics, 10*(21), 2593.
28. Werling, M., et al. (2010). Optimal trajectory generation for dynamic street scenarios in a Frenet frame. In *2010 IEEE International Conference on Robotics and Automation* (pp. 987–993). IEEE.
29. Zhou, B., et al. (2019). Robust and efficient quadrotor trajectory generation for fast autonomous flight. *IEEE Robotics and Automation Letters, 4*(4), 3529–3536.
30. Zhou, X., et al. (2020). Ego-planner: An ESDF-free gradient-based local planner for quadrotors. *IEEE Robotics and Automation Letters, 6*(2), 478–485.

Chapter 8
Drive-by-Wire and Vehicle Control Systems

8.1 Introduction

Drive-by-wire (DbW) technology is a major change in how vehicles are controlled. Instead of using mechanical parts like cables and hydraulic lines to control the engine, brakes, and steering, DbW systems rely on electronic signals. This change is especially important for autonomous vehicles, which need to make fast, accurate, and repeatable decisions without direct human input. With DbW, the physical connection between the driver and the car's mechanical parts is removed. Instead, control is passed through digital systems that work well with software-based control methods. This makes it easier to design and update vehicle systems and improves the precision of vehicle movements.

Modern drive-by-wire systems are essential for the development of smart transportation. They help vehicles work more efficiently, respond quickly to changes in traffic, and communicate with other vehicles and infrastructure through vehicle-to-everything (V2X) networks. DbW systems rely on a combination of sensors, processors, and actuators to read inputs and convert them into precise movements. By moving away from fixed mechanical responses and toward programmable digital control, these systems open the door to better safety, performance, and communication between vehicles. In the following sections, we will look at each component of the drive-by-wire system, how they work together, and why they are important in the larger context of autonomous driving.

Contributor: Zhaofeng Tian

W. Shi, Y. He, *Introduction to Autonomous Driving*,
https://doi.org/10.1007/978-3-031-99485-2_8

8.2 Key Components and Functionality

Drive-by-wire systems are made up of multiple parts that work closely together to control the main functions of a vehicle. These parts replace traditional linkages and improve how quickly and accurately the vehicle responds to inputs. Each component does its own job but also supports the overall feedback loop that keeps the system working reliably.

8.2.1 Electronic Control Unit (ECU)

The Electronic Control Unit, or ECU, is like the brain of the drive-by-wire system. It collects information from sensors and uses that data to send commands to the actuators. Whether those commands come from a human driver or from autonomous software, the ECU ensures that they are carried out in a safe and timely way. Modern ECUs are powerful computers that can run advanced control systems, detect faults, and even predict problems before they happen. The ECU also helps coordinate the timing between sensors and actuators, which is necessary for the vehicle to stay stable and operate smoothly.

8.2.2 Sensors

Sensors give the ECU the information it needs to understand what is happening inside and outside the vehicle. This input is key to making the right control decisions. Different sensors provide different types of information:

- **Throttle Position Sensors**: Measure how open the throttle is, which helps the ECU decide how much power to send to the engine.
- **Steering Angle Sensors**: Track how far the steering wheel has turned. This helps the ECU manage lane-keeping and steering corrections.
- **Wheel Speed Sensors**: Monitor how fast each wheel is turning. These are used for traction control and maintaining vehicle stability.
- **LIDAR, Radar, and Cameras**: These sensors help the vehicle "see" its surroundings. They detect things like other vehicles, road markings, and pedestrians. Combining their data gives a full picture of the driving environment.

All of these sensors work together to give the ECU a clear, up-to-date view of the vehicle's condition and its environment. This makes it possible to safely control the vehicle whether it is being driven by a human or by software.

8.2.3 Actuators

Actuators are the parts that carry out the commands from the ECU. They make physical changes to the vehicle based on the digital instructions they receive:

- **Electronic Power Steering (EPS)**: Uses electric motors to help turn the wheels. This replaces older hydraulic systems and allows more flexible and efficient steering control.
- **Electronic Throttle Control (ETC)**: Controls the throttle electronically instead of using a cable. This lets the ECU adjust engine power more smoothly and precisely.
- **Electromechanical Braking Systems**: These handle braking electronically. They can apply the brakes faster and more accurately and work with features like regenerative braking in electric cars.

Actuators are the final step in the control process. Their job is to turn the ECU's digital decisions into real actions like turning, speeding up, or slowing down. Their speed and accuracy directly affect how the vehicle feels and how safe it is to drive.

8.2.4 Communication Networks

Communication networks connect all the sensors, ECUs, and actuators so they can share data quickly and reliably. Common systems include the Controller Area Network (CAN) and automotive Ethernet. These networks are built to work even in tough conditions and are designed to keep data moving with very little delay.

The network ensures that commands and data are delivered in the right order and at the right time. This is especially important for safety, since any delay or lost signal could lead to mistakes in how the car behaves. A well-designed communication system keeps all the parts of the DbW system working in sync.

The full drive-by-wire system includes several specialized subsystems, each focused on a specific vehicle function:

- **Throttle-by-Wire**: Replaces the mechanical throttle cable with an electronic system. The ECU decides how much to open the throttle based on speed, load, and driving mode.
- **Brake-by-Wire**: Controls the brakes electronically instead of using hydraulic lines. This leads to quicker braking responses and allows the system to combine traditional braking with energy recovery in electric vehicles.
- **Steer-by-Wire**: Removes the direct link between the steering wheel and the wheels. This enables features like variable steering sensitivity and better support for self-driving systems.
- **Shift-by-Wire**: Changes gears using electronic signals. This allows smoother gear shifts, easier integration with autonomous controls, and fewer mechanical parts.

These subsystems all work through the communication network and under the coordination of the ECU. Together, they create a complete system that can control the vehicle's movement with high accuracy and flexibility. Their combined performance is what makes modern autonomous and assisted driving possible.

8.3 Advantages and Challenges

Drive-by-wire systems provide several important benefits, especially for autonomous and semi-autonomous vehicles. These systems form the technological foundation for modern vehicle automation by enabling precise, flexible, and software-driven control of core driving functions. As automotive technology evolves, the advantages of DbW systems support a shift toward safer, more energy-efficient, and customizable vehicle behavior. However, the adoption of drive-by-wire also introduces complex technical and regulatory challenges that must be carefully addressed to ensure reliability, safety, and public trust.

8.3.1 Advantages

One of the most important advantages of drive-by-wire systems is the increased precision and responsiveness they bring to vehicle control. Traditional mechanical systems have natural limitations due to physical friction, wear, and delay in transmission. In contrast, electronic systems allow for high-resolution, real-time control of braking, steering, and acceleration. This precise actuation enables smoother driving dynamics, better handling in critical situations, and the seamless operation of advanced features such as lane centering and automated parking.

Drive-by-wire also contributes to significant weight reduction in vehicle architecture. Removing heavy mechanical linkages such as the steering column, hydraulic lines, and metal rods lowers the overall mass of the vehicle. This not only simplifies manufacturing and improves design flexibility, but also has a direct impact on energy efficiency. A lighter vehicle demands less fuel in internal combustion engines and consumes less battery power in electric vehicles, leading to extended driving range and reduced emissions.

Another key advantage lies in the software-defined nature of drive-by-wire systems. Because vehicle behavior is controlled through code, manufacturers can deploy updates to improve performance, introduce new driving modes, or patch safety vulnerabilities without requiring physical changes to the car. This capability supports over-the-air updates, reduces service costs, and keeps vehicles aligned with evolving technology standards. It also allows customization for different user preferences or driving environments.

From a safety standpoint, drive-by-wire systems make it easier to integrate with advanced driver-assistance systems (ADAS). Features like adaptive cruise control, automatic emergency braking, and lane-keeping assist rely on fast, reliable actuation of steering and braking systems. DbW enables the precise and coordinated execution of these features, improving the vehicle's ability to avoid collisions, maintain safe distances, and respond appropriately to changing traffic conditions. As a result, these systems not only improve driver convenience but also reduce the likelihood of human error.

Most importantly, drive-by-wire systems are essential for fully autonomous driving. Autonomous vehicles rely on continuous feedback and rapid control adjustments to navigate complex environments. By enabling direct, programmable control of all major vehicle functions, DbW creates the interface that connects artificial intelligence to the physical behavior of the car. Without such a system, real-time decisions from self-driving algorithms could not be executed with the required accuracy or speed.

8.3.2 Challenges

Despite these compelling benefits, several challenges complicate the widespread deployment of drive-by-wire systems. One of the main concerns is ensuring system reliability. Unlike mechanical systems, which often provide direct physical feedback and natural fail-safes, electronic systems require careful design to prevent single points of failure. If a control unit or data line fails, it could lead to the loss of a critical function like steering or braking. To prevent this, manufacturers must design redundant systems that include backup sensors, processors, and actuators capable of taking over in the event of failure.

Cybersecurity is another pressing challenge. As vehicles become increasingly connected—both to the cloud and to other vehicles—they also become targets for malicious interference. An attacker who gains access to the control system of a car could potentially disrupt or take over its operation. Therefore, robust security measures such as encrypted data channels, real-time threat detection, and secure boot protocols must be integrated into every level of the drive-by-wire architecture.

Maintaining performance under strict real-time requirements is also a major hurdle. To ensure safe and comfortable driving, the system must process large amounts of sensor data, interpret vehicle and environmental conditions, and issue control commands within milliseconds. Any delay or inconsistency in this loop can result in uncomfortable vehicle behavior or dangerous situations. Achieving this level of responsiveness requires not only high-speed processors and fast communication networks but also carefully tuned software that can prioritize tasks and manage timing under all driving conditions.

Regulatory compliance adds another layer of complexity. Drive-by-wire systems must meet international standards for automotive safety, such as ISO 26262, which governs functional safety in road vehicles. These standards require rigorous validation processes, including failure mode analysis, hazard assessment, and extensive testing under diverse conditions. Meeting these requirements increases development time and cost but is necessary to ensure that drive-by-wire systems are safe and reliable for public use.

8.4 Vehicle Control in Autonomous Driving

Vehicle control in autonomous driving involves the real-time coordination of drive-by-wire systems to execute driving maneuvers both safely and efficiently. This coordination allows autonomous systems to interact dynamically with the environment, adjusting speed, steering, and braking in response to road conditions, obstacles, and traffic behavior. Control strategies used in this context are generally categorized into longitudinal control, lateral control, and integrated control, each addressing a specific dimension of vehicle movement.

8.4.1 Longitudinal Control

Longitudinal control governs the forward and backward motion of the vehicle. It focuses on managing acceleration and braking to achieve tasks such as maintaining a set speed, following a lead vehicle at a safe distance, coming to a smooth stop at traffic lights, or avoiding rear-end collisions. Two widely used approaches for longitudinal control are Proportional-Integral-Derivative (PID) control and Model Predictive Control (MPC).

PID control is a classical technique that adjusts vehicle acceleration based on the error between the actual and desired speed. It combines three components—proportional, which reacts to current error; integral, which accounts for past errors; and derivative, which predicts future error trends. When tuned correctly, PID control provides smooth speed regulation and quick responsiveness to speed changes [1].

Model Predictive Control (MPC), on the other hand, takes a more forward-looking approach. It uses a model of vehicle dynamics to predict future behavior over a short time horizon. Based on these predictions, it selects control inputs that optimize a predefined cost function—often balancing ride comfort, energy efficiency, and safety. MPC is particularly useful in complex environments where the vehicle must make subtle trade-offs between competing objectives [2].

8.4.2 Lateral Control

Lateral control is responsible for adjusting the steering angle to maintain the vehicle's position within a lane, follow a planned path, or execute maneuvers such as lane changes and turns. Several algorithms have been developed for this task, each offering different advantages depending on vehicle speed, road curvature, and desired precision.

The Pure Pursuit method is a simple geometric approach that calculates the required steering angle based on a look-ahead point along the planned trajectory. This method is intuitive and works well for lower-speed scenarios but may be less accurate on sharp curves or at high speeds [3].

The Stanley Controller improves upon this by minimizing two types of errors: the distance from the vehicle to the path (cross-track error) and the difference in orientation (heading error). This makes it well suited for highway driving and has been successfully used in real-world autonomous systems [4].

Another advanced method is the Linear Quadratic Regulator (LQR), which formulates the control problem as an optimization that balances minimizing steering error against the effort required to achieve that correction. LQR offers stability and control smoothness, particularly in dynamic driving conditions [5].

Emerging research has also focused on neural network-based controllers. These leverage data from past experiences to learn how to make steering decisions in complex and uncertain environments. Unlike classical controllers, they can adapt to changing conditions such as varying road surfaces, sensor noise, or unexpected obstacles, thereby improving robustness [6, 7].

8.4.3 Integrated Control Strategies

While longitudinal and lateral control are often designed independently, real-world driving requires them to function together. Integrated control strategies aim to coordinate both types of control simultaneously to ensure smooth, safe, and efficient vehicle behavior. This is especially important in complex scenarios such as navigating roundabouts, merging onto highways, or performing emergency avoidance maneuvers.

Integrated control systems use models of the full vehicle dynamics and consider interactions between acceleration, braking, and steering when making control decisions. For example, when performing a sharp lane change while accelerating, the system must account for how changes in speed will affect lateral stability and vice versa. By synchronizing these functions, integrated strategies improve overall performance and passenger comfort, and they reduce the likelihood of unsafe actions during dynamic maneuvers.

8.5 PID Control for Autonomous Vehicle Control

PID (Proportional-Integral-Derivative) control is a fundamental feedback control method widely used in autonomous vehicle systems to regulate dynamic behavior such as speed, steering, and position tracking. This form of control is based on the continuous monitoring of the difference, or error, between a target value (such as a desired speed or lane position) and the actual state of the vehicle. The controller adjusts its output in real time to reduce this error. This makes PID control especially useful for tasks that require precise tracking or stability under changing environmental conditions, such as driving through curves, maintaining consistent speed in traffic, or staying centered in a lane.

The general form of the PID control law is expressed as:

$$u(t) = K_p e(t) + K_i \int_0^t e(\tau)\, d\tau + K_d \frac{de(t)}{dt} \tag{8.1}$$

In this equation, $u(t)$ represents the control signal that is sent to the vehicle actuators—such as throttle, brake force, or steering angle. The term $e(t)$ is the error, defined as the difference between the desired state and the actual measured state at time t. The constants K_p, K_i, and K_d are the proportional, integral, and derivative gains, respectively. These gains determine how strongly the controller reacts to current, accumulated, and predicted errors.

Each term in the PID equation serves a specific role in shaping the system's response. The proportional term, $K_p e(t)$, generates a control effort that is directly proportional to the current error. This component is responsible for providing an immediate correction whenever a deviation from the desired value is detected. A high proportional gain results in a stronger correction, leading to faster response, but if set too high, it can also cause the system to overshoot the target and oscillate, reducing stability. Thus, careful tuning is necessary to balance quick error correction with smooth operation.

The integral term, $K_i \int_0^t e(\tau)\, d\tau$, addresses accumulated error over time. Even if the proportional response is functioning, small persistent errors might remain—this is known as steady-state error. The integral term adds up these small errors to gradually eliminate them, ensuring the system converges exactly to the desired state. However, excessive use of the integral term can make the system sluggish or unstable, especially in scenarios where external conditions change rapidly, so it must be tuned with caution.

The derivative term, $K_d \frac{de(t)}{dt}$, predicts future error behavior by observing how the error is changing. If the error is increasing rapidly, the derivative term reacts by applying a counteracting force to dampen this trend. This anticipatory behavior helps prevent overshoot and reduces the likelihood of oscillations, particularly in systems exposed to sudden disturbances or high-frequency noise. In autonomous vehicles, the derivative term contributes significantly to maintaining comfort and smoothness during maneuvers.

When combined, the proportional, integral, and derivative terms create a controller capable of balancing immediate responsiveness, long-term accuracy, and predictive stability. This makes the PID controller suitable for various control tasks in autonomous vehicles. For instance, it can manage throttle inputs to maintain a set cruising speed, modulate brake pressure during car-following scenarios, or adjust steering angles to ensure the vehicle remains centered within its lane. The ability to tune the gains K_p, K_i, and K_d provides engineers with flexibility to customize the behavior of the controller for different use cases, road conditions, and vehicle dynamics.

Despite its many strengths, PID control also has limitations. It assumes that the system being controlled is relatively linear and that its dynamics do not change significantly over time. In highly dynamic, nonlinear, or uncertain environments—such as dense urban traffic or slippery road surfaces—PID may not perform optimally without adaptive tuning or additional layers of logic. Nevertheless, its ease of implementation, transparency, and effectiveness under stable conditions ensure that PID control remains a central tool in the control architecture of autonomous vehicles, often forming the baseline upon which more advanced or adaptive methods are built.

8.6 Model Predictive Control for Autonomous Vehicle Control

Model Predictive Control (MPC) is an advanced, optimization-based control technique that has become increasingly important in the domain of autonomous vehicle control. Unlike PID control, which reacts to current and past errors, MPC anticipates future behavior by leveraging a predictive model of the vehicle's dynamics. It uses this model to forecast the system's future states over a specified time horizon and solves an optimization problem to determine the control inputs that best achieve a given objective while satisfying system constraints.

The MPC optimization problem is typically framed to minimize a cost function that penalizes deviation from a desired trajectory and excessive control effort. A common form of the cost function is:

$$\min_{U} \sum_{k=0}^{N} \left[x_k^T Q x_k + u_k^T R u_k \right] \tag{8.2}$$

In this formulation, x_k represents the system state vector at the k-th prediction step, encompassing variables such as position, orientation, and velocity of the vehicle. The term u_k denotes the control input vector at the same step, typically consisting of throttle, braking, and steering commands. The matrices Q and R are user-defined weighting matrices that assign relative importance to errors in state variables and the magnitude of control efforts, respectively. A well-chosen Q places more emphasis on minimizing deviations from the desired path or speed, while an appropriate R helps to avoid aggressive or erratic control inputs.

The states x_k and inputs u_k are interconnected through the system dynamics, which are modeled using state-space equations:

$$x_{k+1} = Ax_k + Bu_k \tag{8.3}$$

$$y_k = Cx_k \tag{8.4}$$

Here, the matrix A captures how the current state evolves over time in the absence of control, B encodes the influence of the control inputs on the state transition, and C maps internal states to measurable outputs. These dynamics ensure that the predicted future states follow a physically plausible trajectory based on the current vehicle behavior and the applied inputs.

Constraints are applied to both states and inputs to maintain realistic and safe operation. For example, the bounds $x_{min} \leq x_k \leq x_{max}$ may enforce limitations on vehicle speed, yaw rate, or lateral position, while $u_{min} \leq u_k \leq u_{max}$ ensures that actuators do not exceed physical limits such as maximum braking force or steering angle. These constraints are crucial for preventing unsafe maneuvers and for ensuring that the optimization respects vehicle capabilities.

MPC operates through a receding horizon control strategy. At each control interval, it predicts a sequence of future states and control inputs over the finite horizon N, solves the optimization problem to minimize the cost function subject to the system dynamics and constraints, and then implements only the first control action. The system state is then updated with new sensor data, and the optimization is solved again in the next step. This cycle repeats continuously, allowing the controller to remain responsive to real-time changes in the environment or vehicle state.

In autonomous driving, MPC is especially effective for trajectory tracking, where the goal is to follow a pre-defined path as closely as possible. It also supports speed and acceleration control by adjusting inputs to maintain comfort, energy efficiency, and regulatory compliance. In more complex scenarios, such as navigating through dynamic environments or avoiding collisions, MPC can evaluate multiple future scenarios, weigh their outcomes, and select the control sequence that best meets safety and performance goals.

The relationship among the different variables is central to how MPC functions. The states x_k describe where the vehicle is and how it is moving. The control inputs u_k are the means through which the controller influences these states. The dynamic model $x_{k+1} = Ax_k + Bu_k$ ties together past states and control inputs to predict future behavior. The cost function uses x_k and u_k to evaluate how well a given control plan meets the objectives, while the constraints on x_k and u_k keep the plan within safe and feasible limits. The optimization problem thus becomes a tightly integrated framework where all variables—states, inputs, and dynamics—work together to generate intelligent, future-oriented control actions.

A key strength of MPC is its ability to handle multi-objective control problems. By tuning the cost function weights in Q and R, the controller can be configured to prioritize different objectives depending on the driving scenario. For example, in dense urban traffic, comfort and collision avoidance may be prioritized, while on highways, efficiency and lane-keeping might take precedence. This flexibility makes MPC a powerful tool for managing the complex and variable demands of autonomous driving.

8.7 Future Trends and Conclusion

The future development of drive-by-wire systems is inseparable from the continued evolution of autonomous vehicle technology. As vehicles become more software-driven and increasingly reliant on artificial intelligence, the design and capabilities of DbW systems will need to evolve to meet higher standards of performance, adaptability, and safety.

One significant trend is the growing adoption of steer-by-wire systems. By fully removing the mechanical linkages between the steering wheel and the wheels, steer-by-wire technology allows for more flexible vehicle design. This enables new interior layouts, including reconfigurable cabins and collapsible steering mechanisms in autonomous vehicles. Furthermore, steer-by-wire opens the door to customizable steering characteristics, such as dynamic feedback tuning based on driving conditions, which can enhance both safety and user experience.

Another important direction is the application of artificial intelligence to optimize vehicle control. Machine learning algorithms, particularly those trained on large datasets of driving behavior and environmental interactions, will enable more adaptive and robust control strategies. These AI-based systems can learn from past experience, predict future events, and adjust control outputs accordingly. This allows autonomous vehicles to operate more safely in complex or unpredictable environments, such as urban areas with erratic traffic patterns or during adverse weather conditions.

To meet the demands of functional safety and reliability, future drive-by-wire systems will incorporate enhanced redundancy and fault-tolerant design. This includes the use of multiple parallel control units, independent power supplies, and diagnostic systems capable of detecting and responding to faults in real time. Redundant communication paths and actuator systems will ensure that the vehicle retains basic control functionality even in the presence of hardware or software failures. As safety standards continue to evolve, these features will be essential for meeting both regulatory requirements and public expectations.

In conclusion, drive-by-wire technology stands as a cornerstone of the next generation of autonomous and intelligent vehicles. Its ability to deliver precise, flexible, and software-defined control is fundamental to enabling advanced automation features. As new innovations emerge in AI, system design, and safety assurance, drive-by-wire systems will become increasingly sophisticated, resilient, and central to the future of mobility.

8.8 Hands-on Exercises

8.8.1 *Steering Control via Raspberry Pi*

Objective This exercise introduces students to the fundamentals of drive-by-wire steering systems by using a Raspberry Pi to send digital steering commands. The goal is to simulate electronic steering by controlling a servo or motor to achieve basic turn angles such as 30° left, 30° right, and center alignment.

Task Students will build a steering setup using a Raspberry Pi and a motor controller such as an L298N or PCA9685. They will write Python code to control the actuator and test commands for left, center, and right turning angles. The sequence of motion should follow: turn left to 30°, return to center, turn right to 30°, and return to center again. Optionally, students may add a feedback mechanism such as a potentiometer to verify angular output. The exercise should demonstrate an understanding of how digital control replaces mechanical linkage in modern vehicles.

Deliverables Each student will submit a short written report, no more than two pages in length. The report must include a diagram of the system wiring, key code excerpts showing how commands were issued, a summary of observed outcomes, and a brief reflection linking the exercise to drive-by-wire principles from Chap. 8, particularly Sects. 8.2–8.4. Students will also submit their complete Python source code. Optionally, they may include a video or photograph of the system performing the programmed turning sequence.

Recommended Papers to Read

1. Ang, K. H., Chong, G., & Li, Y. (2005). PID control system analysis, design, and technology. *IEEE Transactions on Control Systems Technology, 13*(4), 559–576. [1]
2. Wang, C., & Hill, D. J. (2006). Learning from neural control. *IEEE Transactions on Neural Networks, 17*(1), 130–146. [7]

References

1. Ang, K. H., Chong, G., & Li, Y. (2005). PID control system analysis, design, and technology. *IEEE Transactions on Control Systems Technology, 13*(4), 559–576.
2. Ji, J., et al. (2016). Path planning and tracking for vehicle collision avoidance based on model predictive control with multiconstraints. *IEEE Transactions on Vehicular Technology, 66*(2), 952–964.
3. Samuel, M., Hussein, M., & Mohamad, M. B. (2016). A review of some pure-pursuit based path tracking techniques for control of autonomous vehicle. *International Journal of Computer Applications, 135*(1), 35–38.
4. Thrun, S., et al. (2006). Stanley: The robot that won the DARPA grand challenge. *Journal of Field Robotics, 23*(9), 661–692.

5. Scokaert, P. O. M., & Rawlings, J. B. (1998). Constrained linear quadratic regulation. In *IEEE Transactions on Automatic Control, 43*(8), 1163–1169.
6. Sugeno, M. (1985). An introductory survey of fuzzy control. *Information Sciences, 36*(1–2), 59–83.
7. Wang, C., & Hill, D. J. (2006). Learning from neural control. *IEEE Transactions on Neural Networks, 17*(1), 130–146.

Chapter 9
Computing Systems

9.1 Introduction

Autonomous vehicles (AVs) represent the convergence of several advanced technologies, combining robotics, artificial intelligence, control theory, and embedded systems engineering into a cohesive, real-time decision-making platform. The computational demands of AVs far exceed those of conventional vehicles, requiring the seamless orchestration of heterogeneous software and hardware systems to operate safely and reliably in dynamic environments.

Unlike human drivers, who continuously perceive, reason, and act in real-time, AVs must emulate this cognitive loop using a structured computational pipeline. This pipeline includes distinct yet interdependent stages: sensor data acquisition, perception and scene understanding, localization within a mapped environment, motion planning, and vehicular control. Each of these stages imposes strict latency and reliability requirements on the underlying computing infrastructure. Moreover, the pipeline must be capable of operating continuously, in parallel, and in real-world conditions marked by uncertainty and complexity.

At the heart of AV software development is the middleware layer, which abstracts hardware communication, facilitates modularity, and enables distributed system design. Middleware frameworks like Robot Operating System 2 (ROS 2) provide essential services such as message passing, time synchronization, and node orchestration, which are critical for maintaining consistent operation across software components developed by different teams or vendors. The modularity afforded by ROS 2 also supports real-time updates, debugging, and testing, which are vital for the safety-critical nature of AV systems.

Real-time performance is not simply desirable—it is essential. This chapter addresses the constraints and design principles behind Real-Time Operating Systems (RTOS), which guarantee bounded execution times and deterministic task scheduling. These systems ensure that perception or control tasks do not suffer from unpredictable latencies that could compromise vehicle safety.

W. Shi, Y. He, *Introduction to Autonomous Driving*,
https://doi.org/10.1007/978-3-031-99485-2_9

Another central topic is the vehicle programming interface, which defines how software modules communicate with the vehicle's physical components—steering, braking, throttle, and drive-by-wire systems. A robust programming interface ensures that high-level decisions made in the computational pipeline translate into precise physical actions with millisecond-level precision.

Finally, the chapter delves into hardware requirements. AVs rely on specialized computing platforms capable of supporting massive parallelism, fast memory access, and hardware redundancy. This includes high-performance GPUs and TPUs for AI workloads, microcontrollers for low-level actuation, and dedicated safety processors to monitor system health and recover from faults.

9.2 Computational Pipeline

The computational pipeline of an autonomous vehicle is a structured, multi-stage processing architecture that enables the vehicle to perceive its environment, make informed decisions, and act with precision in real time. This pipeline transforms heterogeneous sensor inputs into actionable control signals through a cascade of specialized computational modules. Each stage in this pipeline must execute within strict time constraints and with high reliability to support safe autonomous driving in complex, unpredictable environments [6].

The first stage, sensor data acquisition, involves collecting data from a diverse set of on-board sensors, including high-resolution cameras, LiDAR, radar, ultrasonic transducers, and GPS/IMU units. These sensors offer complementary perspectives: cameras provide rich visual details, LiDAR supplies 3D spatial information, radar performs well in adverse weather, and GPS/IMU supports global positioning and inertial tracking. Sensor data acquisition requires precise time synchronization, calibration, and interface abstraction to ensure that all downstream modules operate on temporally and spatially aligned information [7].

Sensor data is typically acquired and digitized by embedded microcontrollers or dedicated processors located on or near each sensor module. These local processors handle initial tasks such as analog-to-digital conversion, error detection, formatting, timestamping, and in some cases, early-stage filtering or compression. Once prepared, the data is transmitted to the central computing system over a vehicle's internal communication networks [11].

The five major communication interfaces used for this purpose are Controller Area Network (CAN), Universal Serial Bus (USB), Gigabit Multimedia Serial Link (GMSL), Peripheral Component Interconnect Express (PCIe), and automotive-grade Ethernet. Each has distinct characteristics in terms of bandwidth, latency, reliability, and application domain.

CAN is a robust, low-bandwidth (typically up to 1 Mbps in classical CAN, and 5–8 Mbps in CAN FD) protocol well-suited for transmitting small control messages with high fault tolerance. It is typically used for low-frequency sensors such as ultrasonic range finders and system health monitors.

USB, especially USB 3.0 and USB 3.1, can support bandwidths ranging from 5 to 10 Gbps. It is often used for camera modules in prototype vehicles or lower-end AV platforms due to ease of integration. However, USB is susceptible to EMI and lacks automotive-grade robustness, making it less favorable for high-reliability deployments.

GMSL is a high-speed serial link protocol developed specifically for automotive environments. It offers data rates of up to 12 Gbps per link and allows the transmission of video, control, and power over a single coaxial cable. GMSL is particularly suited for high-resolution, low-latency camera feeds in safety-critical perception systems.

PCIe provides extremely high bandwidth (ranging from 8 Gbps in PCIe Gen 3 per lane to over 64 Gbps in Gen 5 $\times 8$ configurations) with minimal latency, and is typically used for integrating LiDARs, high-throughput radars, and high-frame-rate cameras directly with the AV compute core. It supports direct memory access (DMA), allowing rapid data offload to GPUs or other accelerators.

Automotive Ethernet, particularly versions supporting 100 Mbps (100BASE-T1), 1 Gbps (1000BASE-T1), and emerging 10 Gbps (10GBASE-T1) standards, is increasingly deployed for mid- to high-bandwidth sensors. It offers scalable, EMI-resistant, full-duplex communication over twisted pair wiring with support for time-sensitive networking (TSN), making it ideal for synchronized multi-sensor systems.

To achieve deterministic fusion, data transmission is synchronized using time protocols like IEEE 1588 Precision Time Protocol (PTP), ensuring sub-microsecond alignment across all sensor inputs.

Despite these interfaces, performance bottlenecks often arise within the I/O subsystems of the central computing platform. These include oversubscribed bus architectures, limited memory bandwidth when shuttling data between system memory and accelerators, and contention in cache hierarchies. Unoptimized DMA configurations, driver inefficiencies, and resource conflicts across concurrent processes can degrade latency and throughput. Thus, careful mapping of sensors to the appropriate bus, strategic placement of edge computing nodes, and proactive memory management are essential to sustain real-time operations in AV pipelines [6] (Table 9.1).

Following acquisition, preprocessing and sensor fusion are employed to clean, normalize, and merge sensor data into a unified representation of the vehicle's surroundings. Preprocessing tasks such as denoising, spatial filtering, and temporal smoothing are typically executed on the central processing unit (CPU), although certain hardware accelerators or field-programmable gate arrays (FPGAs) may be used for high-throughput, low-latency tasks. Sensor fusion, depending on the complexity and parallelizability of the algorithm, may run on either the CPU or the graphics processing unit (GPU). Classical methods such as Kalman or Bayesian filters are generally CPU-bound due to their sequential nature, while neural fusion networks, which involve matrix operations and convolutional layers, are more efficiently handled by GPUs [9]. Some AV systems also incorporate digital signal processors (DSPs) and specialized AI chips (e.g., TPUs or NPUs) for optimized

Table 9.1 Comparison of sensor communication protocols in autonomous vehicles

Interface	Typical bandwidth	Latency	Robustness	Best for
CAN	1 Mbps (CAN)	Low	High (fault-tolerant)	Control signals, ultrasonic sensors
USB 3.0/3.1	5–10 Gbps	Moderate	Low (EMI-sensitive)	Prototype cameras
GMSL	Up to 12 Gbps per link	Low	High (automotive-grade)	High-res, low-latency cameras
PCIe Gen 3/5	8–64 Gbps	Very Low	Moderate (requires shielding)	LiDAR, radar, high-speed cameras
Ethernet	100 Mbps–10 Gbps	Low	High (automotive-grade)	Synchronized multi-sensor systems

low-power, high-speed execution of specific workloads. Data is staged in RAM for immediate access and often streamed directly from buffer memory to the processing unit through direct memory access (DMA), minimizing CPU overhead and reducing latency.

The memory subsystem plays a critical supporting role by buffering incoming data streams and enabling efficient access patterns during computation. However, bottlenecks in this stage often arise from memory bandwidth limitations and bus contention, especially when multiple high-throughput sensors are active simultaneously. Additionally, inefficient memory allocation or non-optimized data structures can increase latency and reduce throughput, emphasizing the need for architecture-aware software design [11].

The next stage, perception and scene understanding, leverages computer vision and machine learning to extract semantic information from the fused sensor data. This includes object detection and classification, lane and road boundary recognition, traffic sign interpretation, and motion tracking of dynamic agents. These operations are computationally intensive and are most effectively executed on GPUs, which are optimized for parallel processing and deep learning workloads. Certain models may also utilize specialized AI accelerators such as Tensor Processing Units (TPUs) or Neural Processing Units (NPUs) for low-latency inference. CPUs are still involved in orchestrating task scheduling, memory management, and handling non-parallelizable operations. Outputs from this stage form a high-level representation of the environment known as the semantic map, which is then passed on to downstream modules for localization and planning.

Localization follows as a critical module for determining the vehicle's precise position and orientation within a global or local coordinate frame. High-accuracy localization is typically achieved by integrating GPS signals with visual or LiDAR-based map matching techniques. Simultaneous Localization and Mapping (SLAM) may be employed in GPS-denied environments. Localization algorithms, particularly those involving particle filters, visual odometry, or SLAM, are predominantly executed on the CPU due to their iterative and data-structured nature. However, visual-inertial odometry or map-matching methods that involve convolutional neural networks (CNNs) or deep learning components may offload portions of the workload to the GPU [8]. The memory subsystem supports continuous access to large maps and sensor logs, and in high-precision systems, solid-state drives (SSDs) or NVMe storage are used to cache high-definition map tiles in real time. Bottlenecks in localization often arise from limited memory bandwidth, delays in accessing map data, or from synchronization mismatches between asynchronous sensor inputs such as GPS, IMU, and camera data. Robust synchronization, efficient memory caching, and hybrid CPU-GPU parallelization are critical to maintaining localization accuracy and timeliness.

Prediction and planning are responsible for forecasting the future behavior of surrounding agents and computing a safe and efficient path for the AV. Prediction models, often based on recurrent neural networks or probabilistic models, are computationally demanding and typically executed on GPUs or specialized AI accelerators due to their reliance on temporal sequence processing and matrix operations. Planning algorithms, depending on their structure, may run on either the CPU or GPU. Algorithms like A* or RRT* tend to be CPU-bound due to their recursive or graph traversal characteristics, while sampling-based or optimization-heavy methods such as model predictive planning benefit from GPU acceleration.

Bottlenecks in this stage often emerge from the real-time constraints placed on multi-agent forecasting, high-dimensional path search spaces, and the need to integrate constraints such as dynamic obstacles and traffic rules. Computational delays can result from inefficient memory access patterns or from contention between prediction and planning threads sharing GPU resources. Load balancing between heterogeneous processors and optimized scheduling are critical to sustaining low-latency, high-reliability performance in this decision-making phase [9].

Finally, control systems translate the planned path into low-level actuator commands for steering, throttle, and braking. These systems must ensure stability and responsiveness in various driving conditions. Control algorithms include classical methods like PID controllers as well as more advanced techniques like Model Predictive Control (MPC) that consider vehicle dynamics and constraints in the optimization process. These algorithms are primarily executed on real-time capable CPUs or microcontrollers that are directly interfaced with the vehicle's drive-by-wire system. In high-performance setups, control computations may also leverage digital signal processors (DSPs) or dedicated embedded control units to meet sub-millisecond execution requirements. Bottlenecks in this stage often stem from communication delays between the central computing unit and actuator controllers, or from jitter in real-time task scheduling. Ensuring deterministic behavior through real-time operating systems (RTOS), minimal I/O latency, and priority-based thread management is essential for maintaining control stability [11].

The pipeline must be supported by a robust computational infrastructure capable of real-time execution and inter-process communication. Tasks are distributed across CPUs, GPUs, and specialized accelerators, with careful scheduling and priority management. Middleware frameworks, introduced in the next section, are essential to coordinating the flow of data and control signals throughout the pipeline [10].

In sum, the AV computational pipeline is an end-to-end processing architecture where each component builds upon the outputs of the previous stages. A breakdown or delay in any one module can jeopardize the entire decision-making loop, making efficient and resilient design a fundamental requirement [6].

9.3 Middleware Frameworks

9.3.1 Purpose and Role of Middleware

Middleware frameworks are essential for managing the complexity and modularity of autonomous vehicle (AV) software systems. They function as a software abstraction layer that sits between the application layer (perception, planning, control, etc.) and the underlying hardware (sensors, actuators, and computing platforms). This architectural separation allows developers to implement, test, and update high-level algorithms independently of hardware specifics, significantly reducing integration complexity. Middleware facilitates communication, data exchange, and service coordination among distributed software modules, avoiding brittle, tightly coupled architectures [10]. Middleware also provides a unified interface to hardware, abstracting away the nuances of different sensor and actuator protocols, thereby enabling more portable and maintainable code.

To ensure the safe and efficient functioning of AVs, middleware must support real-time performance, deterministic communication, fault tolerance, scalability, and modularity. Real-time performance ensures that system deadlines are met, especially for latency-sensitive tasks like collision avoidance or emergency braking. Deterministic communication guarantees predictable data delivery, which is crucial in safety-critical systems. Middleware must also support fault-tolerant operation by detecting and isolating failures, thereby maintaining overall system stability. Scalability is necessary to accommodate growing numbers of sensors and computing units, while modularity allows software to evolve without extensive re-engineering. Without middleware, system components would require custom interfaces and synchronization logic, increasing software fragility and development complexity [11].

9.3.2 Core Features and Communication Models

At the core of a middleware framework are mechanisms for inter-process communication (IPC), which enable physically or logically separated software modules—such as perception, localization, or planning—to exchange data in real time. This design promotes a separation of concerns where each module can be developed and tested independently, improving code quality and reuse. Moreover, IPC ensures that system upgrades or patches can be deployed with minimal disruption to other modules. IPC mechanisms simplify integration, debugging, and scalability by removing the need for tightly coupled code, which is difficult to maintain in large, distributed systems.

Among IPC paradigms, the publish-subscribe (pub-sub) model is especially effective in AV systems. This model decouples data producers (publishers) from data consumers (subscribers), allowing asynchronous operation and dynamic scalability. Publishers broadcast messages to one or more topics, while subscribers independently register interest in relevant topics. This loose coupling supports system extensibility, as new components can be added without modifying existing ones. Pub-sub enables sensor data, control commands, or map updates to be broadcast to multiple components without requiring knowledge of downstream logic—a necessity when modules like path planning and behavior prediction simultaneously rely on perception outputs.

Beyond communication, middleware frameworks manage serialization, converting in-memory data into formats suitable for transmission. Serialization ensures data consistency and portability across heterogeneous computing environments. For AVs, this means converting large data structures—such as multi-dimensional point clouds or high-resolution images—into compact, transmittable formats. Efficient serialization must strike a balance between compression and decoding latency to ensure timely data delivery [9]. Middleware may implement custom serialization formats optimized for specific sensor modalities to reduce transmission and processing overhead.

Accurate time synchronization is another indispensable feature. AVs rely on tightly coordinated sensor fusion, achievable only when all sensor inputs are timestamped against a common clock. Time synchronization ensures that data collected from multiple sensors at different locations can be accurately correlated in a global reference frame. Middleware often integrates protocols like IEEE 1588 PTP or GNSS-based clocks to guarantee sub-millisecond alignment across distributed components [5]. Inconsistent timing can lead to fusion errors, degraded perception, or unstable control behavior, making robust synchronization mechanisms a foundational requirement.

Resource discovery enables dynamic detection and registration of services or data channels, enhancing system flexibility. This feature allows the middleware to identify and integrate new components at runtime, such as replacing a malfunctioning sensor or integrating a new perception module during development. Resource discovery reduces system downtime and simplifies maintenance by enabling plug-and-play compatibility.

Lifecycle management further contributes to system robustness by defining operational states—startup, initialization, active, idle, or shutdown—and ensuring safe transitions. It manages module dependencies, ensuring that components are activated or deactivated in a controlled sequence. Lifecycle awareness facilitates diagnostics, failure recovery, and rolling updates by enabling selective restarts of malfunctioning components without bringing down the entire system.

Together, these capabilities—IPC, pub-sub, serialization, synchronization, resource discovery, and lifecycle management—form the operational backbone for enabling real-time, reliable, and scalable coordination of AV functionalities. Middleware ensures that complex AV systems remain manageable, adaptable, and robust across a range of operational scenarios.

9.3.3 Open and Standardized Middleware

ROS 2 is an open-source middleware widely adopted in AV development, built on the DDS (Data Distribution Service) standard. It supports a modular architecture through node-based communication primitives like topics, services, and actions. This design facilitates the decomposition of AV functionalities into reusable components and promotes rapid prototyping. The extensive set of developer tools and libraries provided by ROS 2 further accelerates development, especially in academic and pre-commercial research settings.

While DDS offers configurable Quality of Service (QoS) policies for latency and reliability, ROS 2 does not natively guarantee hard real-time behavior. Its default implementations may behave non-deterministically under high load or due to thread scheduling unpredictability. This unpredictability arises from reliance on general-purpose operating systems and user-space process management. Therefore, ROS 2 is not suitable for critical, low-latency control loops without additional engineering efforts [11]. Integration with real-time operating systems or use of commercially hardened variants such as Apex.OS is required to meet bounded execution time and safety constraints [1]. These commercial solutions provide real-time task scheduling, deterministic memory access, and safety certification for deployment in road-ready AVs.

Although flexible, ROS 2's generalized architecture—intended for broad robotic applications—can introduce abstraction layers and runtime overhead, which may impede performance in latency-critical or high-throughput AV environments. Its abstraction over DDS introduces variability in transport latency and requires careful tuning to maintain system responsiveness. Further customization is often needed to ensure compliance with strict safety and timing requirements, including deterministic message delivery and bounded jitter. Moreover, ROS 2 lacks native support for ISO 26262 functional safety standards, necessitating additional wrappers or specialized commercial versions for use in production-grade AVs. These safety enhancements often include static analysis tools, formal verification, and failover mechanisms to meet automotive-grade reliability.

Despite these limitations, ROS 2 remains foundational in AV research, prototyping, and simulation. Its extensibility, large developer community, and ecosystem of tools provide a fertile ground for experimental development and rapid iteration. For academic environments and early-stage startups, ROS 2 lowers the barrier to entry for AV software development and enables testing of novel algorithms and architectures without high upfront infrastructure costs.

Beyond traditional middleware such as ROS and its successor ROS 2, several systems have emerged to address the increasingly critical demand for real-time capabilities in autonomous driving. These include Cyber, developed within the Apollo project, and ERDOS, originating from research in dynamic deadline-driven execution models. Cyber replaces conventional threads with coroutines and manages scheduling entirely in user space, reducing the overhead and latency associated with kernel-space task switching. Processors within Cyber execute coroutines from priority queues, supporting both classic and choreography scheduling schemes to isolate critical tasks. In contrast, ERDOS builds on the D3 (Dynamic Deadline-Driven) model, wherein each operator in the system pipeline is assigned a dynamically adjusted deadline. The runtime then manages execution to maximize the quality of outputs before those deadlines are reached. Both systems reflect the trend of middleware not merely acting as communication substrates, but actively managing real-time performance across distributed computation graphs.

In addition, seL4 has emerged as a high-assurance microkernel with formally verified isolation properties. While not middleware in the traditional sense, seL4 can underpin middleware frameworks by enforcing strict access control and determinism at the OS level. Its support for domain scheduling and fixed-priority preemptive threads enables stronger guarantees for time-sensitive tasks. These middleware and OS-level innovations collectively reflect the increasing emphasis on system-wide coordination for temporal predictability in autonomous driving stacks.

9.4 Real Time Requirements

Autonomous vehicles must respond to sensor inputs and control events with minimal latency and guaranteed predictability. This requirement makes a Real-Time Operating System (RTOS) a critical component in the software stack of an AV. An RTOS differs from a general-purpose operating system in that it guarantees bounded response times for task execution, scheduling, and interrupt handling—an essential feature for the time-sensitive demands of vehicle control systems [1].

In an AV, real-time requirements are most stringent in low-level control loops (e.g., steering, throttle, braking), safety monitors, and actuator coordination. These functions must execute at deterministic frequencies and respond to input changes within strict timing constraints to ensure safe and stable vehicle behavior. For example, a delay in executing an emergency braking routine due to CPU contention or unbounded thread latency could result in a collision. Real-time responsiveness is also crucial for dynamic path planning and collision avoidance where the system must quickly evaluate multiple scenarios and implement maneuvers within a narrow time window.

Meeting real-time requirements in autonomous driving is a multi-faceted challenge. While a common heuristic such as the 100ms end-to-end latency bound is often cited, real deployments must navigate more complex tradeoffs involving task dependencies, communication overhead, and asynchronous data flow. Contemporary systems often employ directed acyclic graph (DAG) models to describe task execution, with either multi-rate periodic triggering or event-driven processing chains. These models help structure the system's real-time analysis but diverge in their assumptions and implications. For example, multi-rate DAGs—common in chassis control systems—assume fixed periodic activations, enabling predictable behavior but risking data overwrites when sampling rates diverge. Conversely, processing chain DAGs, more prevalent in high-level decision pipelines, use active dependencies to trigger tasks upon message arrival, offering lower latency but requiring careful management of message queues and scheduling.

An RTOS provides a deterministic kernel that supports priority-based preemptive scheduling, inter-process synchronization mechanisms, and hardware abstraction layers. It manages system resources in such a way that high-priority tasks can always preempt lower-priority tasks and meet deadlines. It also enables fine-grained control over memory allocation, often forbidding or tightly regulating dynamic memory usage to prevent fragmentation and timing unpredictability. Additionally, most RTOS implementations include static configuration tools and analyzers to verify real-time constraints prior to deployment, which is particularly useful during certification for safety-critical domains.

Real-time operating systems used in AVs include commercial RTOS solutions such as QNX, VxWorks, INTEGRITY, and open-source variants like RTEMS or PREEMPT-RT (a real-time extension of the Linux kernel). These RTOS platforms are typically certified to ISO 26262 standards, particularly at Automotive Safety Integrity Level D (ASIL-D), and are capable of running on embedded control units (ECUs) dedicated to safety-critical functions. QNX and INTEGRITY, for instance, offer certified kernels with support for memory partitioning, process isolation, and temporal determinism—features that enhance both safety and security guarantees for autonomous vehicle platforms.

The integration of an RTOS into the AV software stack often follows a tiered approach. Safety-critical components run on a dedicated RTOS-managed domain, isolated from non-critical components that may operate under a standard Linux or embedded POSIX system. Hypervisors or microkernel architectures are sometimes employed to enforce this separation at the hardware level. These architectures provide virtualized domains that isolate workloads, allowing simultaneous execution of safety-critical and infotainment software while minimizing fault propagation.

AV developers must tune the real-time system carefully—profiling execution paths, minimizing interrupt latency, setting correct task priorities, and validating time budgets under load—to ensure end-to-end determinism. Developers often use real-time tracing and performance profiling tools to ensure that timing budgets are met across heterogeneous workloads. Middleware platforms and real-time frameworks must also align with the RTOS's timing model to avoid jitter and missed deadlines. For example, synchronization primitives and memory access patterns used by middleware should conform to real-time design principles to maintain timing guarantees.

Furthermore, the diversity in middleware and OS-level designs imposes significant variance in how real-time guarantees are approached. ROS 2, while widely adopted, suffers from priority inversion due to its wait-set update mechanics and static callback priority rules. In response, research systems like ERDOS propose cooperative schedulers with application-level deadline assignment, and Cyber shifts the scheduling granularity to coroutines, decoupling data acquisition from thread context switches. These alternative paradigms illustrate a broader recognition that achieving real-time guarantees in autonomous vehicles cannot be isolated to individual components—it must be a system-level property coordinated across software layers, communication infrastructure, and operating system support.

In sum, an RTOS is indispensable for ensuring that the safety-critical parts of an AV execute reliably and predictably, no matter the complexity or variability of the operating environment. It is the foundation on which responsive, certifiable, and safe autonomy is built. Without a reliable RTOS foundation, the assurance of meeting hard deadlines in critical modules—such as emergency braking or evasive maneuvering—would be fundamentally compromised.

A critical link between the software stack and the physical motion of an autonomous vehicle is the **vehicle programming interface (VPI)**. This interface acts as a software-defined bridge that connects high-level decisions made by perception, planning, and control modules to the actual actuation of vehicle hardware such as steering, throttle, brake, and transmission systems. It defines how commands are structured, transmitted, and acknowledged in real time, and how feedback from the vehicle is processed by upstream software modules [12].

In a typical AV architecture, high-level control outputs such as desired velocity, angular rate, or trajectory points are converted into actuator-level commands through a control interface that often includes drive-by-wire systems. These systems replace traditional mechanical linkages with electronic signals, enabling software to exert precise and programmable influence over the vehicle's behavior [7]. This architecture improves responsiveness and reduces mechanical complexity, but it also requires rigorous interface management and failsafe design.

The vehicle programming interface must support bidirectional communication. Outbound messages carry control commands such as throttle percentage or brake pressure, while inbound messages provide continuous feedback including wheel speeds, yaw rate, steering angle, and diagnostic information. These interfaces are typically implemented over `CAN` (Controller Area Network), FlexRay, or Ethernet-based protocols, chosen based on trade-offs among bandwidth, latency, redundancy, and fault tolerance [4]. Ensuring robustness across these links is essential for achieving the deterministic performance expected in safety-critical applications.

To ensure safety and integrity, the interface layer often includes watchdog mechanisms, command arbitration logic, and fallback routines. These features detect anomalies (e.g., command timeout, invalid actuator state) and trigger predefined responses such as gradual deceleration or safe-state reversion. Functional safety standards like `ISO 26262` guide the design and validation of these control paths, emphasizing error detection, redundancy, isolation, and traceability [2].

Modern VPI designs are evolving to accommodate the needs of software-defined vehicles (SDVs) and heterogeneous vehicle computing (VC) platforms. A recent architectural framework defines VPI in terms of five functional categories:

1. **Hardware Interface**, which abstracts sensors, actuators, and communication peripherals.
2. **Data**, which standardizes access to real-time and historical vehicle data.
3. **Computation**, which provides software hooks into execution environments like CPUs, GPUs, and accelerators.
4. **Service**, which enables API-level access to services such as diagnostics, OTA updates, and user-defined applications.
5. **Management**, which controls resource allocation, scheduling, and system lifecycle.

These abstractions allow application developers to interface with vehicle internals without needing to understand low-level hardware or vendor-specific configurations. As a result, VPI simplifies cross-platform development, promotes reusability, facilitates modularity, and provides a critical foundation for scalable AV software deployment [12].

The *OpenVDAP* framework (Open Vehicle-Data Abstraction Platform) is a reference implementation that validates the VPI concept [13]. It integrates a microkernel-based software stack that provides dynamic service discovery, virtualization, and fine-grained control over VPI components. Compared to traditional systems such as *AUTOSAR* or domain-specific APIs, VPI offers more flexibility and modular extensibility while still aligning with safety standards [3].

The paper by Wu et al. [12] introduces a comprehensive vehicle computing model that positions VPI as the connective tissue within a larger platformized vehicle software ecosystem. In this model, VPI is not limited to command relaying; it becomes a unifying abstraction across disparate compute nodes and heterogeneous execution environments. The authors emphasize the importance of abstracting hardware-software boundaries to enable real-time control for safety-critical functions while concurrently supporting lower-priority services like infotainment and diagnostics.

Their VPI model integrates resource-aware task scheduling, domain isolation, and real-time control delegation. This provides a framework for scalable deployment and certifiability. Wu et al. further propose a taxonomy of VPI interactions organized by computation modality (edge, central, hybrid), security domain (trusted vs. untrusted), and data flow type (control vs. data plane). This structured approach clarifies how AV software components interact across shared platforms while maintaining performance and safety boundaries.

Wu et al. also demonstrate the role of VPI in over-the-air (OTA) update frameworks. Here, VPI enables modular and safe updates by managing interface consistency and isolating critical services. Techniques such as service sandboxing and state versioning ensure that new software can be deployed without interrupting ongoing control processes, enabling continuous integration and deployment without compromising safety.

Middleware and API layers often encapsulate the complexities of VPI, allowing algorithm designers to remain hardware-agnostic. These abstraction layers also support simulation environments, enabling software-in-the-loop testing before hardware integration. Through simulation, developers can validate VPI logic, detect edge-case faults, and measure real-time performance under diverse scenarios.

In summary, the vehicle programming interface serves as the execution endpoint of autonomy: it translates software intent into physical action. As autonomous vehicles grow in complexity, their software ecosystems span a wide array of hardware vendors, operating systems, and computational topologies. VPI becomes essential by decoupling software logic from hardware intricacies, enabling AV developers to concentrate on high-level functionality.

Without VPI, software would be tightly bound to hardware-specific APIs and communication protocols, leading to high integration costs, reduced portability, and barriers to certification. Instead, VPI standardizes command semantics, enables modular certification paths, and facilitates platform interoperability. This abstraction supports continuous deployment and scalable development—two key demands of modern AV engineering.

Additionally, VPI promotes orchestration across heterogeneous compute platforms. As AV stacks integrate high-performance GPUs, domain-specific accelerators, and embedded controllers, VPI offers a unified contract for managing distributed control logic. The OpenVDAP study illustrates this interoperability, showing how perception, planning, and actuation logic can be ported across architectures while preserving timing and control fidelity.

From a safety and certifiability perspective, VPI enforces strict module boundaries and supports interface-level validation. By isolating control logic behind standardized interfaces, it becomes possible to test, certify, and iterate on modules independently, in line with `ISO 26262` principles. This design pattern ensures that safety cases can be constructed incrementally, supporting reuse and evolution in AV deployments.

Ultimately, VPI is more than an interface layer—it is a keystone in the architecture of software-defined autonomous vehicles. It supports scalability, maintains system robustness, and enables the transition from experimental platforms to production-grade autonomy that is certifiable, maintainable, and future-ready.

9.5 Hardware Requirements for Autonomous Vehicles

The computational demands of autonomous vehicles (AVs) are among the most intensive in the embedded systems domain. To support real-time perception, planning, localization, and control, the hardware architecture of an AV must integrate high-performance computing platforms, sensor fusion interfaces, data storage systems, and robust power and thermal management—all within the constraints of automotive reliability and safety. These requirements go beyond typical embedded systems used in consumer electronics or industrial control, as AVs must operate with low latency, high availability, and deterministic behavior in dynamically changing, safety-critical environments.

At the core of AV hardware are high-performance processors capable of executing parallel workloads at low latency. These include multicore CPUs for executing sequential and logic-heavy tasks such as decision-making and state estimation, GPUs for accelerating parallelizable workloads such as convolutional neural networks, and dedicated hardware accelerators such as TPUs (Tensor Processing Units), NPUs (Neural Processing Units), and FPGAs (Field-Programmable Gate Arrays) for customized, high-efficiency computation. These diverse processing units are typically organized into a heterogeneous compute cluster located in centralized compute boxes or distributed across real-time control ECUs and sensor-edge devices. Each unit is selected based on its ability to execute specific AV tasks within defined power, latency, and thermal envelopes.

AVs are equipped with a rich array of sensors—cameras, LiDARs, radars, ultrasonic transducers, and IMUs—that generate gigabytes of data per second. This high-bandwidth, multi-modal sensor data must be transferred from peripheral interfaces to the central compute platform with minimal latency and packet loss. To meet these constraints, AV platforms incorporate high-speed data buses such as automotive Ethernet (1–10 Gbps), PCIe (up to 64 Gbps in modern systems), and GMSL (Gigabit Multimedia Serial Link) for camera data streaming. These are often coupled with Direct Memory Access (DMA) engines to bypass CPU bottlenecks and enable direct sensor-to-memory data paths. The entire pipeline must ensure timestamp accuracy, minimal jitter, and synchronization to maintain temporal alignment across sensor modalities.

Memory architecture is another critical element in AV hardware design. The systems require large and fast memory buffers to accommodate various types of data processing pipelines. Incoming sensor frames from high-resolution cameras and LiDARs demand rapid and high-bandwidth writes to DRAM. Intermediate feature maps generated by convolutional neural networks during perception require substantial temporary storage and quick retrieval across multiple cores. Localization modules need fast access to stored map data and historical trajectories to calculate precise vehicle position.

To meet these needs, high-bandwidth DRAM technologies such as LPDDR5 and GDDR6 are employed to support fast, parallel memory access by CPUs, GPUs, and accelerators. NVMe-based solid-state drives are used for persistent storage of high-definition maps, log data from test and deployment environments, and state snapshots that are useful for failure recovery and debugging. These drives must support high IOPS and throughput to avoid bottlenecks during large data reads or writes.

The interaction between cache (e.g., L1/L2/L3), DRAM, and NVM must be carefully tuned to minimize latency, avoid memory contention, and ensure coherency across heterogeneous processing units. This involves fine-grained memory allocation strategies, cache-coherence protocols, and memory scheduling algorithms. Shared memory access models are implemented where multiple modules need access to common datasets, such as fused sensor grids. Memory protection mechanisms, such as address space partitioning and privilege-level enforcement, are essential to isolate critical real-time tasks from non-critical or experimental modules, thereby maintaining predictable performance and system stability.

A key enabler in this space is NVIDIA's Jetson platform, which provides a unified memory architecture (UMA) designed for automotive edge intelligence. UMA allows CPUs and GPUs to share a single, coherent memory space, thereby eliminating the need for explicit data copying between processing elements. This architectural simplification reduces both latency and power consumption, two critical constraints in embedded AV deployments. It also minimizes memory allocation complexity and enables more seamless pipeline execution across heterogeneous cores.

In a typical AV application, vision data captured from cameras is processed through a sequence of GPU-accelerated neural networks for object detection and segmentation. With traditional memory architectures, each stage of the pipeline might require data to be copied across device boundaries, incurring performance penalties. In contrast, UMA enables zero-copy execution, allowing inference models to operate directly on data residing in shared memory, which accelerates throughput and simplifies software development.

Jetson's UMA model also supports cache coherence across cores, ensuring that updates to memory by one processor are immediately visible to others. This feature is essential when integrating real-time control tasks with asynchronous AI inference modules. For instance, a control algorithm running on the CPU can immediately act on the results of a GPU-accelerated perception module without introducing synchronization delays.

This architecture is especially effective in scenarios requiring real-time edge inference, such as low-latency object tracking, adaptive lane detection, free-space estimation, and local decision-making. Jetson devices are frequently deployed at the edge—physically close to the sensor sources—enabling pre-processing, filtering, and prioritization of sensor data before transmission to centralized compute nodes. This distributed intelligence improves overall system scalability and responsiveness.

Thus, NVIDIA's UMA contributes directly to the responsiveness, efficiency, and modularity of autonomous vehicle compute systems. It provides a well-integrated platform for deploying sophisticated AI workloads under real-time constraints, making it a practical solution for embedded intelligence at the automotive edge.

Power distribution and thermal management are also essential design concerns. High-performance processors and memory systems consume significant power and generate considerable heat, especially during peak processing intervals such as object detection or route re-planning. To maintain continuous operation, AV hardware employs dynamic power scaling, thermal-aware task scheduling, and real-time telemetry for monitoring component health. Passive and active cooling systems—such as heat sinks, fans, and liquid-cooled enclosures—are deployed to regulate system temperature. In safety-critical contexts, backup power supplies, redundant power rails, and thermal cutoff mechanisms are included to guarantee continued operation or graceful degradation during fault conditions.

Hardware platforms must comply with functional safety standards such as ISO 26262. This means that every processing path handling safety-relevant functions must be developed and validated under strict safety processes. Features such as ECC (Error Correction Code) memory protect against soft faults in DRAM, while hardware watchdogs monitor CPU activity and reset the system if it becomes unresponsive. Safety processors may operate in lockstep mode—executing identical instructions in parallel and comparing outputs to detect errors—or in triple modular redundancy configurations, where majority voting logic is used to mitigate hardware failures. Diagnostic coverage metrics quantify the system's ability to detect and react to faults in real time.

In recent years, AV architectures have evolved toward domain and zonal controller topologies. In domain architectures, compute resources are grouped by function—such as perception, planning, or actuation—while in zonal architectures, computation is colocated with physical regions of the vehicle, simplifying wiring and reducing latency. Zonal controllers communicate over high-speed backbones with central domain controllers or cloud gateways. These architectural trends allow easier integration of heterogeneous hardware, reduce system complexity, and support flexible reconfiguration across different vehicle models and configurations.

In summary, AV hardware platforms must achieve a fine balance between computational capability, power efficiency, data throughput, and safety assurance. Each component of the hardware architecture must be selected and integrated with an awareness of real-time constraints, environmental reliability, and long-term maintainability in order to support scalable and certifiable autonomous driving systems. Hardware decisions must also anticipate future scalability needs, including support for new sensor technologies, higher-resolution data streams, and evolving machine learning models.

9.6 Hands-on Exercises

9.6.1 Exercise: Enhancing Ubuntu for Autonomous Driving Deployment

Objective Explore how a standard Ubuntu Linux system can be optimized to serve as the foundation for an autonomous vehicle (AV) computing platform. Students will identify system-level modifications to support real-time constraints, high-throughput data handling, middleware integration, and interface with AV hardware and software stacks.

Task You are given a clean Ubuntu 22.04 LTS installation. Your task is to examine and modify this system to better meet the needs of a real-time autonomous driving application. You should simulate the deployment environment of an edge-based AV computing unit that integrates with sensor fusion, planning, and control modules. Specifically:

Analyze the kernel for real-time performance support. Investigate preempt-RT patches and kernel scheduling behavior.

Set up ROS 2 middleware (e.g., Humble Hawksbill) and assess DDS communication performance under simulated load.

Configure GPU acceleration and CUDA libraries compatible with AV software stacks (e.g., Autoware).

Evaluate I/O performance for LiDAR and camera data (consider DMA buffers, memory-mapped I/O).

Inspect systemd service dependencies and boot time optimization.

Assess security modules and isolation mechanisms (AppArmor/SELinux, namespaces) relevant for multi-container AV deployment.

Deliverables

A report (2–4 pages) detailing:

System-level modifications made to the Ubuntu configuration.

Performance profiling results before and after modifications (e.g., using `latencytop`, `perf`, or `rttest`).

Justification of each change with reference to real-time and AV computing system requirements (as described in Chap. 9: real-time constraints, DDS throughput, etc.).

A working Dockerfile or shell script that automates the deployment of your improved system.

Optional: A benchmark demonstrating improved inter-process communication (e.g., ROS 2 latency test between publishers and subscribers).

Note This exercise simulates preparing a Linux-based computing system for deployment in an edge vehicle. Use Chap. 9's discussion on computational pipelines, middleware, and hardware-software co-design as your technical reference.

Recommended Papers to Read

1. Liu, L., Lu, S., Zhong, R., Wu, B., Yao, Y., Zhang, Q., & Shi, W. (2020). Computing systems for autonomous driving: State of the art and challenges. *IEEE Internet of Things Journal, 8*(8), 6469–6486. [6]
2. Liu, L., Liu, S., & Shi, W. (2021). 4C: A computation, communication, and control co-design framework for CAVs. *IEEE Wireless Communications, 28*(4), 42–48. [7]
3. Wu, T., Wu, B., Wang, S., Liu, L., Liu, S., Bao, Y., & Shi, W. (2021). Oops! it's too late. Your autonomous driving system needs a faster middleware. *IEEE Robotics and Automation Letters, 6*(4), 7301–7308. [11]

References

1. Becker, J., Sagar, M., & Pangercic, D. (2021). A safety-certified vehicle OS to enable software-defined vehicles. In *Automatisiertes Fahren 2021: Vom assistierten zum autonomen Fahren 7. Internationale ATZ-Fachtagung* (pp. 51–67). Springer.
2. Carlsson, D. (2013). Development of an ISO 26262 ASIL D compliant verification system.
3. Fürst, S., & Bechter, M. (2016). AUTOSAR for connected and autonomous vehicles: The AUTOSAR adaptive platform. In *2016 46th annual IEEE/IFIP international conference on dependable systems and networks workshop (DSN-W)* (pp. 215–217). IEEE.
4. Fürst, S., et al. (2009). AUTOSAR–A worldwide standard is on the road. In *14th international VDI congress electronic systems for vehicles, Baden-Baden* (Vol. 62, Chap. 5). Citeseer.
5. Hu, S., Li, G., & Shi, W. (2021). Lars: A latency-aware and real-time scheduling framework for edge-enabled internet of vehicles. *IEEE Transactions on Services Computing, 16*(1), 398–411.
6. Liu, L., et al. (2020). Computing systems for autonomous driving: State of the art and challenges. *IEEE Internet of Things Journal, 8*(8), 6469–6486.
7. Liu, L., Liu, S., & Shi, W. (2021). 4C: A computation, communication, and control co-design framework for CAVs. In *IEEE Wireless Communications, 28*(4), 42–48.
8. Liu, L., Wang, Y., & Shi, W. (2022). Understanding time variations of dnn inference in autonomous driving. Preprint, arXiv:2209.05487.
9. Liu, L., et al. (2022). Prophet: Realizing a predictable real-time perception pipeline for autonomous vehicles. In *2022 IEEE real-time systems symposium (RTSS)* (pp. 305–317). IEEE.
10. Macenski, S., et al. (2022). Robot operating system 2: Design, architecture, and uses in the wild. *Science Robotics, 7*(66), eabm6074.
11. Wu, T., et al. (2021). Oops! it's too late. your autonomous driving system needs a faster middleware. *IEEE Robotics and Automation Letters, 6*(4), 7301–7308.
12. Wu, B.-F., et al. (2024). Vpi: Vehicle programming interface for vehicle computing. *Journal of Computer Science and Technology, 39*(1), 22–44.
13. Zhang, Q., et al. (2018). OpenVDAP: An open vehicular data analytics platform for CAVs. In *2018 IEEE 38th international conference on distributed computing systems (ICDCS)* (pp. 1310–1320). IEEE.

Chapter 10
End-to-End Solutions

10.1 Introduction

End-to-end autonomous driving represents a paradigm shift in the development of intelligent driving systems by replacing traditional modular pipelines with unified neural architectures. These systems aim to learn a direct mapping from sensory inputs to driving actions, enabling more cohesive and data-driven decision-making. By circumventing the need for manually engineered intermediate representations, end-to-end approaches have the potential to simplify the system architecture, reduce error propagation, and adapt more flexibly to complex and dynamic driving environments. However, the design and deployment of such systems come with their own challenges, including interpretability, data efficiency, safety, and generalization. In this chapter, we explore recent advancements in end-to-end autonomous driving, examining the latest models, cooperative frameworks, generative approaches, and surveys that shape the state-of-the-art in this field.

10.2 Recent Research

Recent work in autonomous driving has focused heavily on integrating perception, prediction, and planning into unified frameworks. Hu et al.[3] present UniAD, a planning-oriented architecture that aligns all subsystems toward the final task of motion planning. By embedding perception, prediction, and planning as interrelated components within a shared network, UniAD reduces the compounding errors typical of sequential pipelines and achieves state-of-the-art performance on the nuScenes benchmark.

Xu et al. [6] extend end-to-end driving systems by integrating large language models into a multimodal framework. Their model, DriveGPT4, processes multi-frame videos and textual queries to output interpretable driving decisions. This

W. Shi, Y. He, *Introduction to Autonomous Driving*,
https://doi.org/10.1007/978-3-031-99485-2_10

approach improves transparency and facilitates query-based reasoning, making it a significant step toward explainable autonomous driving systems.

A major limitation of traditional sensor-only autonomous driving is addressed by Yu et al.[7], who propose UniV2X, a vehicle-to-everything (V2X) cooperative end-to-end driving system. Their hybrid dense-sparse transmission and fusion pipeline enables enhanced situational awareness and planning via infrastructure cooperation. UniV2X unifies agent perception, occupancy prediction, and planning in a single framework, demonstrating strong improvements over purely onboard approaches.

Zheng et al.[8] reformulate autonomous driving as a generative task in GenAD, which employs a variational autoencoder to learn trajectory priors in a structured latent space. Their instance-centric tokenization method captures rich scene context, allowing simultaneous planning and prediction through temporal modeling. This model achieves state-of-the-art performance on vision-centric planning tasks.

Multimodal sensor fusion is explored in depth by Prakash et al.[4], who develop TransFuser, a transformer-based model that integrates camera and LiDAR data using attention mechanisms. This method demonstrates significant improvements in urban navigation, especially in complex scenes requiring global reasoning and long-term planning.

To contextualize these algorithmic developments, Chen et al. [1] offer a comprehensive survey of over 270 works on end-to-end driving. They categorize approaches based on architecture, challenges, and emerging trends, highlighting the growing importance of integrated and interpretable systems.

The benefits of multimodal fusion are further analyzed by Xiao et al. [5], who examine different RGB-D fusion strategies. They show that combining color and depth modalities enhances generalization and robustness, particularly in challenging lighting and environmental conditions.

Finally, Chib and Singh[2] provide a deep-learning-focused survey covering the entire AV pipeline from perception to control. Their work highlights the main algorithmic trends and challenges, including scalability, data efficiency, and real-world deployment considerations, making it a useful reference for future research directions.

10.3 Advantages and Limitations of End-to-End Approaches

End-to-end autonomous driving offers several compelling advantages. Chief among these is architectural simplification: by collapsing perception, planning, and control into a single trainable model, such systems avoid error accumulation inherent in traditional modular stacks. This integration allows for holistic optimization, potentially resulting in more efficient, reactive, and adaptive decision-making. Additionally, end-to-end models can learn directly from raw sensor data, enabling the discovery of task-relevant features that may be overlooked by manually engineered intermediate representations. This often leads to performance gains in highly dynamic or visually complex environments (Table 10.1).

Table 10.1 Summary of pros and cons of end-to-end autonomous driving

Advantages	Limitations
Simplified architecture: fewer modules, reduced interface errors	Lack of interpretability: decisions are harder to trace and validate
Joint optimization of perception, planning, and control	Requires large, diverse datasets to generalize effectively
Ability to learn task-relevant features directly from raw sensor input	Poor robustness in out-of-distribution or edge-case scenarios
Better adaptation to new environments with retraining or domain adaptation	Challenging to verify and validate due to tightly coupled components
Reduced reliance on hand-crafted rules or heuristics	Limited trust and regulatory acceptance for safety-critical deployment

End-to-end systems also provide a more natural fit for learning-based frameworks. They allow for gradient-based optimization across the entire decision pipeline, making it easier to incorporate rich supervisory signals, such as driving scores or comfort metrics, into training objectives. Furthermore, such systems are inherently adaptable to new driving conditions and domains when paired with techniques like domain adaptation, continual learning, or reinforcement learning. Their reduced dependency on hand-crafted heuristics and rule-based logic simplifies system integration and maintenance.

However, end-to-end systems also face important limitations. One major challenge is interpretability. Unlike modular pipelines, where each component is individually accessible and explainable, end-to-end models often function as black boxes, making it difficult to diagnose errors or validate decisions. This opacity complicates debugging and undermines user trust and regulatory approval. Another concern is data efficiency; end-to-end systems typically require vast, diverse datasets to generalize effectively, and their performance can degrade sharply in out-of-distribution scenarios. Furthermore, safety guarantees are harder to enforce in tightly coupled architectures where small model perturbations may propagate unpredictably across the decision space.

Additionally, the lack of modularity can hinder robustness in edge cases. Traditional systems can fail gracefully if one module encounters unexpected input, whereas end-to-end models are more likely to produce brittle, hard-to-predict behavior. This makes the verification and validation of end-to-end models more complex and currently limits their acceptance in highly regulated or mission-critical contexts.

Consequently, while end-to-end systems offer elegant and potentially more powerful alternatives to traditional approaches, their deployment in real-world, safety-critical applications must be approached with careful consideration of these trade-offs. Hybrid models that combine end-to-end learning with interpretable or modular components may offer a promising compromise, retaining many of the benefits while addressing some of the practical limitations.

10.4 Modular vs. End-to-End Architectures: A Comparative Analysis

The software architecture for autonomous driving has evolved significantly in recent years. Traditional systems have relied on a modular approach that separates functionality into distinct subsystems. However, with advancements in deep learning and large-scale data processing, end-to-end architectures have gained traction as an alternative that promises unified optimization and potentially better generalization. This section explores both paradigms in detail, comparing their design principles, strengths, and challenges, and discusses emerging hybrid models that aim to combine the best of both worlds.

10.4.1 Modular Autonomy: Structured Engineering

Modular systems divide the driving pipeline into individual components, such as perception, prediction, planning, and control. Each module is designed, trained, and tested independently, often with bespoke datasets and evaluation metrics. This separation allows developers to isolate and improve specific capabilities without affecting the entire system.

Each row in Table 10.2 reflects a critical subsystem in the modular stack. These modules communicate via well-defined interfaces, enabling transparency and modular testing. The granularity also allows performance benchmarking for each capability under varied scenarios.

Table 10.2 Example modular AV pipeline

Module	Functionality
Sensors	Hardware components such as cameras, LiDARs, and radars that collect raw data about the surrounding environment, including depth, velocity, and texture. This data serves as the foundational input for all downstream modules.
Perception	Software algorithms that analyze sensor data to identify and classify objects (vehicles, pedestrians, signs), estimate their position and motion, and recognize lane markings and traffic signals. Outputs are typically structured as object lists or occupancy grids.
Prediction	Takes the perceived object data and forecasts the future positions and possible behaviors of other road users. Predictive accuracy is vital for safe planning in dynamic environments.
Planning	Uses predictions to create a trajectory for the ego vehicle that adheres to traffic laws and optimizes for comfort, safety, and efficiency. Planners balance goals like lane-keeping, merging, and obstacle avoidance.
Control	Converts planned trajectories into low-level commands such as steering angle, throttle, and brake actuation, ensuring smooth and accurate vehicle motion.

Table 10.3 Simplified end-to-end driving pipeline

Stage	Description
Raw sensor input	Continuous data streams from multimodal sensors (e.g., RGB images, LiDAR point clouds) fed directly into a model without manual feature extraction or intermediate representation.
Unified model	A monolithic or multitask neural network architecture, often based on convolutional, recurrent, or transformer layers. It processes and fuses sensory inputs to learn a representation of the environment, policies, and decision-making strategies.
Output	The end-to-end model produces either direct control commands (steering, throttle, brake) or an intermediate representation such as a planned trajectory or future ego-positions. This output is interpreted or fed into a low-level controller.

10.4.2 End-to-End Learning: Unified Optimization

End-to-end models approach autonomous driving as a single learning problem, often mapping sensor inputs directly to control outputs or future trajectories using deep neural networks. This approach simplifies the system architecture and enables holistic optimization over a unified loss function.

Table 10.3 illustrates how the end-to-end architecture condenses perception, prediction, and planning into a single learned function. Unlike modular systems, the network learns to integrate features implicitly across tasks, which can lead to stronger performance on rare or complex interactions.

10.4.3 Trade-off Summary

Table 10.4 provides a comprehensive comparison, contextualizing architectural decisions within practical engineering and operational concerns. Each aspect listed evaluates fundamental trade-offs that impact model development lifecycle, scalability, and societal integration.

10.5 Future Directions: Hybrid and World Model Approaches

While modular and end-to-end systems each offer distinct advantages and face specific limitations, recent research has turned toward hybrid models that aim to integrate the strengths of both. Hybrid autonomy stacks retain structured intermediate representations like those used in modular systems while learning key components with end-to-end deep learning methods. This architectural fusion

Table 10.4 Comparison of modular and end-to-end approaches

Aspect	Modular approach	End-to-end learning
Interpretability	Each module's output can be independently inspected, such as viewing detected objects or planned paths. This supports explainability and trust.	Decisions are encoded within weights and activations of deep networks, which makes inspection and explanation challenging.
Debugging and maintenance	Failures can be traced to a specific subsystem and corrected without retraining others, allowing targeted updates.	Errors are holistic, requiring retraining or modification of the entire model pipeline, which complicates patching and validation.
Training requirements	Demands extensive manual labeling and task-specific data engineering for each module.	Enables training on large-scale unlabelled or semi-labelled datasets using self-supervised or imitation learning.
Generalization	Often rigid due to engineered features and hand-crafted rules, making it brittle under novel conditions.	Learns from diverse data distributions, enhancing robustness to edge cases and out-of-distribution scenarios.
Deployment flexibility	Modules can be updated independently, making incremental deployment and hot-fixes possible.	Requires retraining or end-to-end validation for each change, reducing deployment agility.
Safety and certification	Modules can be tested in isolation using scenario-based and formal methods, facilitating regulatory approval.	Lack of transparency and modularity complicates safety validation and hinders certification under current frameworks.

enables interpretability and modular validation while enhancing data efficiency and generalization.

One compelling avenue is the development of latent world models. These models learn a compact representation of the environment—a latent space—in which forward dynamics can be simulated. Instead of directly predicting control commands, the system imagines potential futures within the latent space and evaluates them to select the most promising plan. Such models are trained to generate future observations or ego-vehicle trajectories conditioned on current and past states. This predictive capacity supports more robust planning under uncertainty and facilitates counterfactual reasoning, where the vehicle can assess hypothetical scenarios (e.g., "what if the pedestrian moves faster?").

In practical terms, world models typically consist of an encoder that maps sensory data into latent representations, a dynamics model that simulates the evolution of the environment in the latent space, and a decoder that reconstructs future observations or values. These components can be implemented using variational autoencoders, recurrent neural networks, or transformer-based architectures.

Planning can then be performed via trajectory sampling, Monte Carlo tree search, or gradient-based optimization within the latent space.

This approach enables simulation-driven training at scale, reducing the reliance on expensive real-world data collection. For example, frameworks like Wayve's GAIA and Google's Dreamer employ imagination-based learning to train agents with limited environmental exposure. These models learn complex, temporally extended behavior policies and exhibit strong zero-shot generalization to unseen environments.

Hybrid systems also benefit from flexible conditioning. Unlike traditional pipelines, world models can incorporate high-level goals, natural language instructions, or map-based priors directly into their decision-making loop. This opens pathways toward more intuitive human-robot interaction and context-aware planning.

From a system design perspective, hybrid architectures offer a spectrum between fully modular and fully end-to-end designs. A practical implementation might use learned perception and prediction modules within a structured planning-and-control stack, or rely on neural representations for scene understanding while retaining rule-based controllers for execution. This design flexibility allows developers to gradually introduce learned components without compromising system safety or validation workflows.

In summary, hybrid and world model approaches present a promising direction for next-generation autonomous systems. They offer a path toward scalable, adaptive, and certifiable autonomy that can operate in complex, real-world environments. These models aim to unify the clarity of modular systems with the representational richness and generalization power of end-to-end learning, marking a pivotal step in the evolution of autonomous vehicle technology

10.6 Hands-on Exercises

10.6.1 Implement an End-to-End Model in Simulation

Task Develop a complete end-to-end autonomous driving model using camera inputs to generate control commands directly within the BlueICE simulation framework. Utilize a pretrained convolutional neural network (e.g., based on YOLO or CNN-LSTM architectures) or train a simple network using simulated data collected from manual or autopilot driving in CARLA.

Objective Use BlueICE co-simulation with CARLA to simulate a driving environment, including sensors (cameras, LiDAR if desired) and an ego vehicle controlled via an end-to-end policy. You may use the Autoware-Universe integration or a standalone Python ROS 2 node to interpret sensor data and issue vehicle control commands.

Deliverable Submit a short report with the following: (1) a brief description of your model architecture, (2) a description of the training data or pre-trained weights used, (3) video capture or screenshots of the model driving autonomously in BlueICE, and (4) performance metrics such as average lane deviation, number of collisions, or time-to-complete a defined course. Include code and instructions to reproduce the experiment.

10.6.2 Compare and Contrast the Performance of Modular vs End-to-end Approaches

Task Compare the performance of a traditional modular pipeline (e.g., Autoware-based stack with separate perception, planning, and control modules) against the end-to-end approach implemented in Exercise 1.

Objective Use the same test course or scenario within the BlueICE framework for both pipelines. Record data such as localization accuracy, obstacle avoidance behavior, time to completion, control smoothness, and interpretability of system behavior. Include qualitative observations (e.g., responsiveness to dynamic objects or scene changes) and quantitative metrics.

Deliverable Submit a comparative analysis report that includes: (1) a description of both pipelines, (2) tables or plots showing comparative performance metrics, (3) discussion of system robustness, interpretability, and failure modes, and (4) reflections on trade-offs in complexity, modularity, and learning dependency. Include visualizations from simulation logs and annotated screenshots or video clips. Optionally, include a brief proposal for hybrid systems that incorporate strengths from both paradigms.

Recommended Papers to Read

1. Chib, P. S., & Singh, P. (2023). Recent advancements in end-to-end autonomous driving using deep learning: A survey. *IEEE Transactions on Intelligent Vehicles, 9*(1), 103–118. [2]
2. Chen, L., Wu, P., Chitta, K., Jaeger, B., Geiger, A., & Li, H. (2024). End-to-end autonomous driving: Challenges and frontiers. *IEEE Transactions on Pattern Analysis and Machine Intelligence*. [1]
3. Xu, Z., Zhang, Y., Xie, E., Zhao, Z., Guo, Y., Wong, K. Y. K., . . . & Zhao, H. (2024). Drivegpt4: Interpretable end-to-end autonomous driving via large language model. *IEEE Robotics and Automation Letters*. [6]

References

1. Chen, L., et al. (2024). End-to-end autonomous driving: Challenges and frontiers. *IEEE Transactions on Pattern Analysis and Machine Intelligence*, *46*, 10164–10183.
2. Chib, P. S., & Singh, P. (2023). Recent advancements in end-to-end autonomous driving using deep learning: A survey. *IEEE Transactions on Intelligent Vehicles, 9*(1), 103–118.
3. Hu, Y., et al. (2023). Planning-oriented autonomous driving. In *Proceedings of the IEEE/CVF conference on computer vision and pattern recognition* (pp. 17853–17862).
4. Prakash, A., Chitta, K., & Geiger, A. (2021). Multi-modal fusion transformer for end-to-end autonomous driving. In *Proceedings of the IEEE/CVF conference on computer vision and pattern recognition* (pp. 7077–7087).
5. Xiao, Y., et al. (2020). Multimodal end-to-end autonomous driving. *IEEE Transactions on Intelligent Transportation Systems, 23*(1), 537–547.
6. Xu, Z., et al. (2024). Drivegpt4: Interpretable end-to-end autonomous driving via large language model. *IEEE Robotics and Automation Letters*, *9*, 8186–8193.
7. Yu, H., et al. (2025). End-to-end autonomous driving through v2x cooperation. In: *Proceedings of the AAAI conference on artificial intelligence* (Vol. 39, Chap. 9, pp. 9598–9606).
8. Zheng, W., et al. (2024). Genad: Generative end-to-end autonomous driving. In: *European conference on computer vision* (pp. 87–104). Springer.

Chapter 11
Security and Privacy

11.1 Introduction

The rapid advancement of autonomous vehicle technology represents one of the most significant transformations in modern transportation. As we progress towards a future where self-driving vehicles become commonplace, the critical importance of security and privacy in these systems cannot be overstated. This chapter provides a comprehensive examination of the fundamental security and privacy challenges inherent in autonomous vehicle systems.

At their core, AVs represent complex cyber-physical systems, integrating sophisticated hardware components with advanced software algorithms. These vehicles rely on an intricate network of sensors, actuators, and computational units to perceive their environment, make decisions, and execute driving maneuvers. From a computer science perspective, this creates a multifaceted attack surface that demands rigorous security considerations across multiple layers of the system architecture. While conventional security focuses primarily on protecting data and systems from unauthorized access, autonomous vehicle security must simultaneously ensure the physical safety of passengers, pedestrians, and other road users.

The interconnected nature of modern autonomous vehicles further compounds these challenges. AVs operate within a broader ecosystem of intelligent transportation infrastructure, incorporating vehicle-to-everything (V2X) and vehicle-to-vehicle (V2V) communication, vehicle-to-infrastructure (V2I) protocols, and cloud-based services. Each of these communication channels introduces potential vulnerabilities that must be addressed.

This chapter approaches these challenges from both theoretical and practical perspectives. We begin by establishing fundamental concepts in the field of security. Then we delve into state of the art and challenges involving the three aspects

Contributor: Lichen Xia

W. Shi, Y. He, *Introduction to Autonomous Driving*,
https://doi.org/10.1007/978-3-031-99485-2_11

of autonomous driving (Sensors, in-vehicle systems, and V2X) and point out the drawbacks of existing solutions. we will also examine real-world case studies, enabling readers to understand how theoretical security concepts translate into practical attacks and concrete system designs.

The field of AV security and privacy continues to evolve rapidly. As future AV scientists, understanding these challenges and their solutions is crucial for developing the next generation of secure and privacy-preserving autonomous transportation systems.

11.2 The CIA Security Model: Foundation of Information Security

The CIA triad—Confidentiality, Integrity, and Availability—represents the cornerstone principles of information security. In AV systems, as in all critical information systems, these three elements work together to create a comprehensive security framework.

11.2.1 Confidentiality

Confidentiality ensures that information is accessible only to those authorized to view it. Think of confidentiality as a secure vault in a bank—only specific individuals with the proper credentials can access its contents. In autonomous vehicles, confidentiality protects sensitive data such as: Personal information about vehicle occupants, Route histories and destination patterns, System configuration data Authentication credentials and communication between AVs.

11.2.2 Integrity

Integrity guarantees that data remains accurate and unaltered throughout its entire lifecycle. Consider integrity like a tamper-evident seal on a medicine bottle—any unauthorized modification becomes immediately apparent. In the context of autonomous vehicles, integrity is crucial for things like sensor data readings, Control commands and system logs.

11.2.3 Availability

Availability ensures that information and resources are accessible when needed by authorized users. Think of availability as a reliable electrical grid—the power must be there when you need it. In autonomous vehicle systems, availability is critical for tasks such as sensor networks, communication channels, emergency response systems and real-time decision-making processes.

11.3 Attack Surfaces

As autonomous vehicles (AVs) evolve into complex cyber-physical systems, their security posture must be rigorously examined to identify potential vulnerabilities. One of the fundamental principles in cybersecurity is the attack surface, which refers to the sum total of all entry points—both physical and digital—through which an adversary can attempt to exploit a system [23]. In the context of AV security, understanding the attack surface is essential for assessing risk exposure and designing robust defense mechanisms.

Unlike traditional computing environments, where the attack surface is primarily composed of software and network interfaces, AVs present a multi-layered and dynamic attack surface due to their reliance on diverse interconnected components. In this book, we follow the definition of AV'S attack surface from [18]. The attack surface of AV can be broadly categorized into three primary domains: sensors, in-vehicle systems, and vehicle-to-everything (V2X) communication. Each category introduces unique security challenges and potential vulnerabilities. Sensors serve as the "eyes and ears" of an autonomous vehicle, enabling environmental awareness and localization. However, adversaries can exploit weaknesses in sensor technology to manipulate vehicle perception. The internal electronic architecture of an autonomous vehicle consists of interconnected electronic control units (ECUs) and subsystems that manage vehicle operations. Security risks arise from unauthorized access, data tampering, and software exploitation. AVs rely on V2X (Vehicle-to-Everything) communication for real-time interaction with infrastructure, other vehicles, and cloud services. This connectivity expands the attack surface to include network-based threats. Understanding these categories helps in systematically analyzing vulnerabilities within the AV ecosystem and designing targeted security countermeasures. In the remaining of this chapter, for each category, we will talk about key components that can be targeted by attackers and representative attacks.

11.4 Sensor Security

As introduced in Chap. 3, an AV is equipped with a variety of sensors, including, but not limited to, cameras, GNSS/IMU, ultrasonic radar, millimeter-wave radar, and LiDAR. These sensors collect information about the AV's position, its surrounding environment, and other relevant data to help the AV make appropriate decisions in real time. Sensors play a pivotal role in autonomous driving, making them a primary target for cyber and physical attacks. Currently, the majority of attacks on autonomous vehicles exploit sensor vulnerabilities. These attacks typically involve injecting false data or deliberately impairing sensor performance to disrupt vehicle perception. Since different sensors operate based on distinct principles, attackers employ a range of techniques tailored to each sensor type [39]. In this section, we will discuss the existing attacks and potential countermeasures for different types of sensors. For the detail explanation of each sensor, please refer to the content in Chap. 3.

11.4.1 Camera

Autonomous vehicles classified as Level 3 and 4 rely on an intricate network of cameras to perceive their surroundings. These cameras work in unison to detect pedestrians, lane markings, traffic signs, and other vehicles, ensuring safe navigation. One critical aspect of this system is traffic light recognition—when a red light or a pedestrian is detected, the vehicle must slow down or come to a stop to prevent accidents. However, this reliance on visual data also makes autonomous systems vulnerable to manipulation. Malicious actors can exploit these mechanisms by introducing fake traffic lights or artificial pedestrian images, causing the vehicle to halt unnecessarily.

Laser-based Attacks Researchers have demonstrated that malicious actors can exploit the camera's sensitivity by directing a laser beam into the sensor, effectively "blinding" or confusing the camera and, by extension, the driving system [40]. The laser-based attack relies on directing a focused, high-intensity beam at the camera's photodiodes, saturating or overloading them so that the captured image becomes washed out or filled with streaks. In many systems, the automatic gain control (AGC) further compounds this problem by drastically dimming the entire scene in response to the bright spot, rendering the rest of the image underexposed. This combination of local oversaturation and global underexposure creates noise, obscures details, and disrupts critical vision-based functions such as lane detection and object recognition.

Input Attacks Camera-based perception systems form the backbone of many autonomous driving pipelines, enabling visual tasks such as pedestrian detection, lane tracking, and traffic sign recognition. However, these systems are susceptible to adversarial attacks, wherein small but carefully crafted perturbations to images or physical objects cause significant misclassification. [14] showcases how an adversary can systematically manipulate a vehicle's camera input, compromising safety-critical functions without requiring direct access to the vehicle's internal software. The attacker begins by creating an adversarial pattern—often in the form of a sticker, patch, or graphic-optimized through iterative algorithms to exploit vulnerabilities in the vehicle's convolutional neural network (CNN). Once fabricated, this pattern is strategically placed on or near objects in the camera's field of view (e.g., traffic signs, vehicle exteriors, or pedestrians) so that it appears innocuous to human observers. However, the subtle pixel-level perturbations it contains can mislead the CNN into misclassifying or overlooking critical features, effectively neutralizing the camera-based perception system and creating significant hazards for autonomous driving.

Fusion Attacks Modern autonomous vehicles rely on multiple sensor modalities—most commonly cameras and LiDARs—to perceive their surroundings. The camera provides rich color and texture information, while the LiDAR (Light Detection and Ranging) sensor offers accurate depth and distance measurements. By fusing these two data streams, the vehicle's perception system gains robustness against errors or occlusions that might occur if it relied on a single sensor. However, recent research has revealed that adversaries can exploit inherent vulnerabilities in the way camera and LiDAR signals are combined, enabling physical-world attacks that evade detection or inject ghost objects into the fused perception pipeline [6]. Attackers exploit sensor-specific vulnerabilities to create cross-modal mismatches within fusion-based perception systems. By physically crafting adversarial objects—using specialized coatings, shapes, or reflective properties—they trick the camera into misreading color or texture while simultaneously manipulating LiDAR's return signals. This causes the fused data to contain inconsistencies, leading the perception algorithms either to ignore legitimate objects (making them "invisible") or to register phantom obstacles. Because these systems rely on consistent detections across both modalities, even small but strategically placed disruptions can undercut their reliability and ultimately compromise the safety of autonomous vehicles.

Denial-of-Service Attacks Real-time object detection systems running on resource-constrained AV platforms are particularly vulnerable to latency-sensitive attacks. Unlike typical adversarial examples that aim to force misclassification, latency attacks focus on overloading the computation pipeline, causing the detection algorithm to slow down or miss frames altogether. This disruption can lead to delayed or incomplete inferences—potentially disabling safety-critical applications such as pedestrian detection or collision avoidance [8]. The attacker bombards the object detector with inputs designed to be computationally taxing—such as oversized or intricately patterned images—thereby saturating hardware resources and triggering extra processing steps. Because the detection pipeline typically includes operations that become disproportionately expensive when confronted with certain visual complexities, each frame significantly slows down inference. By continuously injecting these demanding inputs in rapid succession, the attacker creates a backlog of unprocessed data, leading to substantial latency spikes, dropped frames, and ultimately degraded or even crippled performance on the edge device.

11.4.2 GNSS/IMU

Global Navigation Satellite System (GNSS) and Inertial Measurement Unit (IMU) technology form a crucial real-time localization method in autonomous driving. Among these, GNSS-RTK (Real-Time Kinematics) stands out for its exceptional accuracy, capable of achieving centimeter-level precision even in dynamic conditions. However, despite its high precision, GNSS-RTK has limitations—it operates at a relatively low update frequency, and its satellite signals can be easily obstructed by buildings, tunnels, or dense foliage.

To compensate for these shortcomings, IMUs and odometers are integrated into the system, tracking displacement and directional changes between successive GNSS-RTK updates. While these sensors provide high-frequency updates, they are susceptible to accumulated errors over time. By combining GNSS with IMU technology, autonomous vehicles can maintain real-time localization with minimal delay, enhanced accuracy, and a consistently high update rate, ensuring reliable navigation even in challenging environments.

Spoofing Attacks GNSS spoofing is one of the most critical attacks on satellite-based navigation [24]. In spoofing, an adversary deliberately broadcasts counterfeit GNSS signals—designed to mimic authentic satellite transmissions—so that a receiver in a target vehicle, drone, or infrastructure system locks onto the fake signals instead of the legitimate ones. By manipulating parameters such as signal strength, timing, or navigational data, the attacker can coerce the victim device to compute incorrect positions or times, potentially causing large-scale navigational failures or system malfunctions. Attackers use software-defined radio (SDR) or specialized signal generators to produce GNSS-like waveforms that mirror the frequencies and modulation schemes of genuine satellites. They carefully synchronize these counterfeit signals in both time and frequency with the legitimate ones, initially broadcasting at lower power to "overlay" and eventually overpower the real signals. Once the victim receiver locks onto these spoofed signals, the attacker manipulates the navigational data—introducing subtle or abrupt shifts in timing and positioning information. As a result, the receiver is tricked into calculating false coordinates or time references while appearing to remain locked onto authentic satellite transmissions.

Jamming Attacks Jamming represents one of the most straightforward yet disruptive methods for undermining GPS-based navigation in autonomous vehicles [50]. In a jamming attack, an adversary floods the GPS frequency bands with overpowering noise or intentionally structured interference. Because GPS signals are already weak by the time they reach Earth's surface, even low-cost jammers can mask or overshadow legitimate satellite broadcasts. When a vehicle's GPS receiver is jammed, it often fails to obtain a valid positional lock, threatening critical functionalities like route planning, geofencing, and real-time localization. Attackers typically use a simple transmitter or software-defined radio (SDR) to emit noise at or near the GPS L1 frequency (1.57542 GHz), flooding the band with signals strong enough to mask genuine satellite broadcasts. Because GPS signals are extremely weak by the time they reach Earth's surface, even a relatively low-powered jammer can overpower them, particularly if placed close to or inside the targeted vehicle. The jammed receiver, swamped by artificial interference, rapidly loses lock on authentic satellite data and can no longer maintain accurate timing or positional information, causing navigation modules to generate errors or resort to fallback systems.

11.4.3 Ultrasonic Sensor

Ultrasonic sensors were initially integrated into vehicles as a key component of automated parking assistance systems. These sensors operate by emitting ultrasonic signals in a specific direction through dedicated transmitters. The moment a signal is sent, a timer is activated. As the ultrasonic waves travel outward, they reflect off obstacles in their path and return to the sensor's receiver. When the reflected signal is detected, the timer stops, and the system calculates the distance between the vehicle and the obstacle based on the elapsed time. This fundamental principle allows vehicles to gauge their surroundings with precision, enhancing safety and maneuverability in tight spaces.

Spoofing Attacks Ultrasonic Sensors are also susceptible to adversarial interference. Specifically, malicious actors can exploit the sensors' reliance on sound wave reflections through two main methods: (1) spoofing, which involves injecting deceptive ultrasonic signals that produce false distance readings; and (2) jamming, wherein noise or overpowered ultrasonic signals saturate the receiver, preventing accurate distance measurements [53, 30, 29]. In a spoofing attack, the adversary transmits ultrasonic pulses at or near the same frequency as the vehicle's sensor, carefully timing and shaping these pulses to generate false echoes that mimic real object reflections. By ensuring the spoofed signals reach the sensor at precise intervals, the attacker tricks the sensor into interpreting the echoes as legitimate obstacles situated at specific distances. The attacker can achieve this with relatively simple and inexpensive hardware, such as an ultrasonic transducer, an amplifier, and a microcontroller, which can cause the vehicle to perceive phantom obstacles or fail to detect real ones, potentially leading to unsafe driving decisions during low-speed maneuvers.

Jamming Attacks In a jamming attack, the adversary blasts the sensor's operating frequency (often around 40kHz) with either broad-spectrum ultrasonic noise or a continuous, high-intensity tone, effectively drowning out the legitimate echoes the sensor relies on. Because the receiver is overwhelmed by this interference, it cannot distinguish the genuine return signals from the noise, causing the vehicle to lose accurate distance measurements. Even moderately powered jammers can achieve this effect, particularly when placed close to or inside the target vehicle, rendering the ultrasonic sensor effectively blind and creating a significant risk of collisions or system errors.

11.4.4 Millimeter-Wave Radar

Millimeter waves refer to electromagnetic waves with wavelengths ranging from 1 to 10 millimeters. In the realm of automotive technology, vehicle-mounted millimeter-wave radar typically operates within the 24 and 77 GHz frequency bands in most countries, though some regions, such as Japan, have also adopted the 60 GHz band. One of the key advantages of millimeter waves is their strong penetration capability, allowing them to function reliably even in adverse weather conditions such as rain, fog, and snow. This resilience makes them an essential component in modern autonomous driving and advanced driver assistance systems, ensuring consistent performance regardless of environmental challenges.

Spoofing Attacks Just like other automotive sensors, Millimeter-Wave Radars can be compromised by malicious interference. Two prominent forms of attack—spoofing and jamming—target the radar's reliance on specific frequency bands and timing measurements, leading to false readings or complete signal disruption [53, 30, 29]. In a spoofing attack, the adversary broadcasts radar-like signals that closely match the vehicle's mmWave radar frequency and modulation. Using software-defined radios (SDRs) and precise timing control, the attacker generates false echoes that arrive at the radar receiver at intervals corresponding to fabricated distances. By adjusting the amplitude and introducing Doppler effects, they can simulate the presence of moving objects, either stationary or approaching. These deceptive signals trick the radar into perceiving non-existent obstacles or misjudging the properties of real ones, potentially resulting in phantom braking, erratic steering, or undetected hazards, especially in high-speed situations requiring accurate collision avoidance.

Jamming Attacks In a jamming attack, the adversary overwhelms the vehicle's mmWave radar by flooding its frequency band with high-power noise, typically using continuous wave (CW) signals or rapidly pulsed interference. This disruption prevents the radar from detecting weak reflections, effectively causing the receiver to misinterpret or miss legitimate echoes. By continuously saturating the radar signal with broad-spectrum or frequency-modulated interference, the attacker renders the radar's signal processing pipeline ineffective, leading to either complete radar failure (blindness) or erratic, unreliable measurements. This can severely impact critical safety functions, such as lane change or collision detection, jeopardizing vehicle control and reaction time.

11.4.5 LiDAR

LiDAR has emerged as one of the most critical sensors in autonomous driving technology. Its operating principle is based on emitting laser beams and capturing the reflected signals from surrounding objects. By analyzing the differences between the emitted and received signals, LiDAR can determine key details such as distance, azimuth, altitude, and even the shape of detected objects. This technology enables the creation of high-definition (HD) maps by capturing dense 3D point cloud data from both stationary and moving elements in the environment. LiDAR's key strengths lie in its long detection range and its exceptional ability to provide highly accurate three-dimensional representations of objects, making it indispensable for precise navigation and obstacle detection in autonomous systems.

Spoofing Attacks LiDAR systems can be deceived by carefully crafted spoofing attacks. In a spoofing attack on LiDAR, the attacker uses a laser emitter to send out counterfeit light pulses synchronized with the victim LiDAR's emission window. These spoofed pulses mimic the expected wavelength and pulse width, and the attacker carefully times their return to simulate objects at specific distances, often at predetermined intervals that correspond to desired ranges. By adjusting pulse intensity, the attacker ensures that the spoofed signals register without being too overwhelming or too faint. This causes the LiDAR to create ghost obstacles or distort real object positions, leading to false object detection or environmental map corruption, which can result in sudden vehicle maneuvers or sensor fusion breakdowns in autonomous driving systems.

Relay Attacks A relay attack on LiDAR leverages the principle of intercepting laser pulses emitted by the sensor and retransmitting them—often with modified timing or intensity—to create 'illusions' of obstacles at false locations or to obscure actual objects in the environment [45, 40]. This attack is particularly concerning for autonomous vehicles that depend on LiDAR data for real-time 3D mapping and obstacle detection. By capturing and forwarding signals in near-real-time, an adversary can manipulate the sensor's perception without necessarily requiring complex spoofing hardware to generate fully synthetic pulses from scratch. In a relay attack on LiDAR, the attacker intercepts the sensor's laser pulses using a photodetector and then retransmits the captured signals, often with modifications in timing, intensity, or angle. By carefully controlling the delay of the relayed pulses, the attacker can create the illusion of objects at false distances, shifting real reflections to simulate nonexistent obstacles or misrepresenting the locations of genuine objects. Additionally, by adjusting parameters like pulse amplitude or divergence, the attacker can saturate the LiDAR's sensor, masking real objects behind a flood of modified echoes. This attack requires minimal processing or spoofing of LiDAR's modulation scheme, relying on the real-time interception and re-emission of pulses to deceive the system.

11.5 System Security

11.5.1 Operating System Security

An AV system consists of multiple interconnected software modules, including localization, perception, planning, and control. These modules must adhere to strict real-time requirements, necessitating an operating system to manage their execution efficiently. The primary role of the operating system is to facilitate communication and allocate resources among these modules.

In this section, we examine the security aspects of the operating system that have been or will be used in AVs. AV sensors continuously generate vast amounts of data during operation, placing significant real-time processing demands on the system. Given the high interdependence of the modules within the autonomous driving system, ensuring seamless communication and efficient resource allocation presents a considerable challenge.

ROS

ROS is a versatile and robust framework for robot programming, built around a distributed multiprocessing architecture that relies on message passing. Many key elements of autonomous driving leverage ROS, including quaternion-based coordinate transformation [55], a 3-D robotic mapping framework [4], and the positioning algorithm SLAM [3]. Its message-driven approach supports a modular software design in which each module can both receive and dispatch messages.

Sensor attacks are external threats that do not need access to the autonomous vehicle's operating system. In contrast, internal attacks occur when someone hacks into the vehicle's operating system. A common security concern with operating systems built on ROS is the lack of authentication for messaging and node creation [25], making them vulnerable to such attacks. There are two main types of attacks [31]: (1) Hijacked ROS Node: An attacker can take control of a ROS node to continuously send out messages. This constant flow of data can overwhelm the system's memory, causing it to run out of memory (OOM). As a result, the autonomous vehicle's operating system may start shutting down ROS node processes, potentially leading to a system crash. (2) Tampered Messages: An attacker can hijack a ROS topic or service to send altered or fake messages. These forged messages can cause the operating system to behave abnormally, compromising the vehicle's functionality.

The first type of attack occurs because ROS lacks an isolation mechanism, allowing any ROS node to access system resources without restrictions. The second attack happens because messages exchanged between nodes are not encrypted, making it easy for attackers to intercept and read their content [33].

SROS [52] is a collection of security upgrades designed for ROS. It includes Transport Layer Security (TLS) to protect communication within ROS and uses x.509 certificates to establish trusted connections. SROS also allows for setting restrictions on ROS nodes through customizable namespaces and assigned roles. Additionally, it provides user-friendly tools to automatically generate node key pairs, audit ROS networks, and create or refine access control policies.

DDS and ROS2

The Data Distribution Service (DDS) [38] was originally used by the U.S. Navy to solve compatibility issues during frequent software upgrades in its complex ship networks [15]. Today, DDS is the standard for publish/subscribe communication in distributed real-time systems. For autonomous vehicles, the operating system must implement a high-speed, efficient DDS framework that works seamlessly across multiple cores, CPUs, and boards. DDS ensures real-time, flexible, and efficient data distribution, meeting the demands of various real-time communication applications. Its security standard uses a three-step handshake process with these messages: (1) HandshakeRequest, (2) HandshakeReply, and (3) HandshakeFinal [26].

The DDS security specification defines five service plugin interfaces (SPIS) to increase security.

1. Authentication Service Plugin: This is the core component of the SPI architecture. It verifies the identity of applications or users performing operations on DDS.
2. Access Control Service Plugin: This plugin sets and enforces rules to control what DDS-related actions a domain participant can perform.
3. Cryptographic Service Plugin: Responsible for all cryptographic tasks, it handles encryption, decryption, hashing, digital signatures, and more.
4. Logging Service Plugin: This plugin enables the auditing of security-related events within DDS, ensuring proper tracking and accountability.
5. Data Tagging Service Plugin: It allows users to attach tags to data samples, making it possible to label and track specific security-related actions within DDS.

Unlike the old master-slave structure, ROS2 uses a more advanced distributed architecture. It relies on DDS (Data Distribution Service) for its messaging system, with built-in security extensions to protect data during transmission [12]. By using DDS, ROS2 enhances the reliability and real-time performance of multi-robot collaboration.

Currently, the security of ROS2 relies heavily on the security features provided by DDS. However, ROS2 only uses the first three Security Plugin Interfaces (SPIs) of DDS mentioned earlier.

1. Builtin Authentication Plugin ("DDS: Auth: PKIDH"): This plugin uses a verified Public Key Infrastructure (PKI). Each participant must have a public key, a private key, and an x.509 certificate for authentication.
2. Builtin Access Control Plugin ("DDS: Access: Permission"): Also based on PKI, this plugin requires each domain participant to have two signed XML documents: Governance file that defines the security policies and permissions file that specifies the participant's allowed actions.

3. Builtin Cryptographic Plugin ("DDS: Crypto: AES-GCM-GMAC"): This plugin ensures secure communication using authenticated encryption with the Advanced Encryption Standard (AES) in Galois Counter Mode (GCM), providing both encryption and data integrity.

However, ROS2 is missing some essential security features. Two key examples are:

1. Secure Over-the-Air (OTA) Updates [22]: OTA updates connect the vehicle manufacturer's server to an autonomous vehicle via Wi-Fi, allowing software updates to be downloaded and installed. However, if hackers compromise this process, the vehicle's security could be at risk.
2. Secure Key Exchange [28]: Current methods for exchanging keys between remote listeners and talkers are not secure enough. This vulnerability can allow attackers to intercept the keys, compromising communication security.

11.5.2 Control System Security

Various mechanical components and digital devices in autonomous vehicles are managed by Electronic Control Units (ECUs). Communication between different ECUs within the vehicle is facilitated through a digital.

CAN

The Controller Area Network (CAN) is the primary communication protocol used in a vehicle's electronic network [9]. It is known for its stability, reliability, real-time performance, strong resistance to interference, and long transmission range. CAN uses differential signal transmission, requiring just two signal lines for normal communication: CAN-H (High) and CAN-L (Low). These two lines carry opposite signals, which helps prevent electromagnetic interference and radiation [37].

In a CAN network, any node can start communication with others at any time, as there is no master-slave structure. However, access to the bus is determined by node priority. Autonomous vehicles often integrate several telematics nodes into the CAN network [51]. These nodes support functions like remote control and software updates. However, hackers can exploit the network by accessing it through the vehicle's onboard diagnostics (OBD) port.

Vulnerabilities of CAN

Currently, the CAN bus lacks authentication and access control, making it an easy target for hackers [54]. Several notable attacks have exploited this vulnerability. For instance, Miller and Valasek [35] used system flaws to remotely access a Jeep's multimedia system. They then targeted the V850 controller, altered its firmware, and gained the ability to send remote commands to the CAN bus, allowing control over critical systems like power and braking. This serious security breach led to the recall of 1.4 million vehicles.

While directly accessing the CAN bus is challenging, certain connected components expose potential entry points. The vehicle's entertainment system and the OBD-II port, used for maintenance, are both linked to the CAN bus and serve as common attack paths. The five most common methods of attacking the CAN bus are outlined below.

1. OBD-II Invasion: The evolution of OBD-II represents a significant advancement in vehicle diagnostics, bringing enhanced functionality and much-needed standardization across the automotive industry. This universal interface—the OBD-II port—serves as the primary gateway for accessing critical vehicle status information. During routine maintenance procedures, automotive technicians rely on proprietary diagnostic software platforms—such as Ford's NGS, Nissan's Consult II, and Toyota's Diagnostic Tester—to interface with the OBD-II port and conduct comprehensive vehicle examinations. However, this technological convenience presents a double-edged sword: since the OBD-II port maintains a direct connection to the vehicle's CAN bus network, malicious actors equipped with appropriate diagnostic software can potentially compromise vehicle security by intercepting CAN bus communications and seizing unauthorized control of vital vehicle systems [27].
2. CD Player Invasion: When the vehicle's media player connects to the CAN bus, it creates a potential security vulnerability. Malicious actors can embed attack code within music CDs that, once played, allows the compromised media player to infiltrate the Controller Area Network. Through this vector, attackers gain unauthorized control over the vehicle's CAN bus system, potentially manipulating critical vehicle functions [7].
3. TPMS Invasion: The TPMS, or tire pressure monitoring system, offers an insidious entry point for malicious actors. Attackers can embed harmful code directly into this seemingly innocuous safety feature. The system, designed to continuously measure air pressure within the tires, becomes weaponized when these dormant instructions recognize specific pressure readings. Once these predetermined thresholds are detected, the concealed malware activates, launching its assault on the vehicle's critical systems [20].
4. Bluetooth Invasion: Autonomous vehicles feature Bluetooth connectivity that enables integration with various electronic devices such as smartphones, personal digital assistants, and laptops. However, this connectivity creates a vulnerability where malicious software installed on smartphones can potentially establish communication with the vehicle's CAN bus system through the Bluetooth connection [32].

As the CAN lacks authentication mechanisms, each transmitted frame merely indicates its intended destination without revealing its source. Consequently, any malicious data presented in the correct message format can be accepted as legitimate by the system, creating a fundamental security vulnerability.

CAN FD

CAN FD maintains full compatibility with standard CAN communication protocols, operating on the same physical layer. This advanced protocol delivers efficient distributed real-time control with robust security features. Each CAN FD node incorporates secure data transmission, sophisticated error detection mechanisms, signaling capabilities, and self-diagnostic functions.

Despite its designation as the next-generation vehicle network protocol, CAN FD inherits certain security vulnerabilities from its predecessor. The broadcast nature of CAN data frames leaves communications susceptible to confidentiality and authentication breaches. Consequently, CAN FD remains vulnerable to eavesdropping and replay attacks, presenting ongoing challenges for automotive network security architects.

11.6 V2X Communication Security

When an autonomous vehicle is on the road, it becomes part of the Internet of Vehicles (IoV). Vehicle-to-Everything (V2X) serves as an umbrella term for the various communication mechanisms within the IoV. As discussed in Sect. 11.1, these mechanisms typically include Vehicle-to-Vehicle (V2V), Vehicle-to-Infrastructure (V2I), Vehicle-to-Pedestrian (V2P), and Vehicle-to-Network (V2N). Through V2X communication, a vehicle can access critical traffic information, such as real-time traffic conditions, pedestrian presence, and the status of surrounding vehicles.

Ensuring the security of V2X communication is a crucial aspect of autonomous driving. In this section, we examine the potential security risks associated with V2X and explore corresponding mitigation strategies.

Within the Vehicle-to-Everything (V2X) architecture, four distinct communication paradigms facilitate the intelligent transportation ecosystem:

Vehicle-to-Vehicle (V2V) communication primarily operates in urban and highway environments, enabling direct data exchange between vehicles. This real-time sharing of critical parameters—including velocity, directional movement, acceleration, braking patterns, relative positioning, and steering data—allows vehicles to anticipate the behavior of nearby traffic participants and implement preemptive safety protocols.

Vehicle-to-Infrastructure (V2I) communication establishes connections between onboard vehicle systems and roadside units (RSUs). These infrastructure components gather data from passing vehicles and disseminate time-sensitive information through connected portals, creating a dynamic information network for all road users.

Vehicle-to-Pedestrian (V2P) communication employs sophisticated sensor arrays to identify and interpret pedestrian behavior in proximity to vehicles. When potentially hazardous situations emerge, vehicles can deploy visual and auditory warnings to alert pedestrians to imminent dangers.

Vehicle-to-Network (V2N) communication links vehicles with cloud-based platforms for comprehensive data exchange. These cloud systems store and process vehicle data to deliver an array of services—from navigation assistance and remote monitoring capabilities to emergency response coordination and entertainment options.

In their influential work, Hasrouny and colleagues [21] developed a systematic framework for categorizing attacks on Vehicle-to-Everything (V2X) communication systems. Their classification organizes these security threats into four fundamental categories based on the specific communication service that is compromised:

Authenticity and Identification Attacks: These threats target the verification processes that ensure vehicles and infrastructure can trust the identity of communication partners. Availability Attacks: These attacks aim to disrupt or deny access to V2X services, preventing legitimate users from utilizing critical communication channels. Data Integrity Attacks: This category encompasses attempts to alter transmitted information without authorization, potentially causing vehicles to make decisions based on falsified data. Confidentiality Attacks: These involve unauthorized access to protected or sensitive information exchanged within the V2X ecosystem.

In the following sections, we will conduct a comprehensive examination of each category, analyzing significant research contributions and real-world implications to provide a thorough understanding of the V2X security landscape.

11.6.1 Authenticity/Identification Attacks

Sybil Attack

In a Vehicular Ad Hoc Network (VANET), each vehicle functions as a wireless node upon joining the network. The dynamic nature of VANETs—where vehicles continuously enter and exit the network—necessitates distributing data across multiple nodes to ensure information remains accessible throughout the network.

This distributed architecture, while enhancing availability, creates security vulnerabilities. A sophisticated attacker can exploit this structure by deploying a single malicious node that assumes multiple virtual identities, a tactic known as a Sybil attack. When data is backed up across the network, multiple copies may unknowingly be stored on the same malicious node masquerading as different entities.

Furthermore, this identity deception enables the propagation of malicious information throughout the network from what appears to be multiple independent sources. Consider a scenario where an attacker disseminates fabricated traffic information to several nodes. When a legitimate vehicle receives identical reports from these seemingly distinct sources, it may consider the information highly credible due to the apparent consensus. Acting on this falsified data, the vehicle might alter its route, potentially leading to dangerous situations or even accidents, as documented by researchers [19].

Key or Certificate Replication Attack

In this type of security breach, an attacker employs network eavesdropping techniques (known as "sniffing") to intercept authentication credentials such as digital certificates or cryptographic keys as they traverse the network. Once these valuable credentials are captured, the attacker exploits them by submitting them to an authentication server. This allows the attacker to successfully establish a seemingly legitimate identity within the system [13].

GNSS Spoofing Attack

In VANETs, the precision and reliability of location data form a critical foundation for the entire network's functionality. Malicious actors can exploit this dependency by creating sophisticated interference systems that generate counterfeit navigation signals. These falsified signals interfere with a vehicle's Global Navigation Satellite System (GNSS) reception, causing it to develop an inaccurate understanding of its position within the environment.

Timing Attack

Timing attacks represent a subtle yet dangerous threat to VANETs by specifically targeting the transmission timing of time-sensitive messages. In these attacks, a malicious node strategically introduces artificial delays to critical communications that have strict real-time requirements.

The consequences of such attacks are particularly severe because many time-critical messages directly impact vehicle safety operations and overall network functionality. For example, emergency brake warnings or collision avoidance alerts become significantly less effective when delayed by even fractions of a second.

As research by scholars has demonstrated [42], these deliberately introduced latencies can substantially compromise both individual vehicle safety and the collective reliability of the entire VANET ecosystem. This vulnerability underscores the importance of not only securing message content but also ensuring the timeliness of message delivery in vehicular communication systems.

11.6.2 Availability Attacks

Denial of Service (DoS) Attack

DoS attack targets VANETs by overwhelming the system with excessive, meaningless requests. This effectively prevents legitimate users from accessing network services as resources become depleted. Such attacks can originate from malicious entities both within and outside the network. When flooded with artificial malicious traffic, legitimate network components like onboard units (OBUs) and roadside units (RSUs) cannot function properly due to resource constraints [56]. The Distributed Denial of Service (DDoS) attack represents a more sophisticated variant of the DoS attack. In this scenario, an attacker gains control over numerous compromised nodes (referred to as "zombie nodes") to launch coordinated DoS attacks against the VANET.

DDoS attacks in VANETs typically manifest in two primary scenarios [17]:

V2V Communication Attacks Zombie nodes transmit message requests to a targeted vehicle from various positions and time intervals. The attacker can modify both the timing and content of these message requests depending on the node type. The goal is to overwhelm the targeted vehicle and disrupt the network, preventing the victim from utilizing network resources.

V2I Communication Attacks These attacks are conducted by vehicles from different locations, targeting roadside units (RSUs). When RSUs become overwhelmed, they cannot respond to legitimate requests from regular network participants.

Spamming Attack

A spamming attack constitutes a specific form of DoS attack within VANETs. In this attack vector, adversaries flood the network with excessive amounts of unsolicited messages (spam), deliberately consuming available bandwidth resources. This consumption of network capacity results in increased transmission latency throughout the Vehicular Ad Hoc Network [48].

The decentralized nature of transmission medium management in VANETs presents a significant challenge for controlling such spamming activities. Without centralized oversight mechanisms, detecting and mitigating these attacks becomes particularly difficult.

Flooding Attack

Flooding attacks represent another variant of DoS attacks targeting vehicular networks. In this attack methodology, malicious actors employ compromised nodes to broadcast fraudulent messages throughout the VANET infrastructure. The propagation of these falsified messages consumes substantial network resources, resulting in diminished network throughput. When a flooding attack succeeds, it can cause temporary network outages, effectively suspending services across the VANET for specific time intervals [43].

Wormhole Attack

In VANETs, a wormhole attack represents a sophisticated security threat wherein malicious entities exploit established private channels within the network infrastructure. Rather than utilizing standard network pathways, these adversaries covertly transmit intercepted information between disparate network locations.

The wormhole creates an illusion that two distant parts of the network are directly connected, disrupting normal routing protocols. Once this artificial tunnel is established, the malicious node gains significant control over the traffic routed through it. This privileged position enables various secondary attacks, including selective packet dropping, unauthorized data modification, comprehensive traffic analysis, and other malicious activities [34].

Blackhole Attack

In the domain of Vehicular Ad Hoc Networks (VANETs), the blackhole attack represents a significant threat to network availability. This attack follows a deceptive two-stage process that exploits the routing mechanisms fundamental to vehicular communications. During the initial route discovery phase, malicious nodes within the network respond to routing request packets with fraudulent claims. These nodes falsely advertise themselves as optimal forwarding candidates, presenting artificially low latency metrics to destination nodes. This deception effectively manipulates the route selection algorithms employed by legitimate nodes.

Consequently, numerous network participants mistakenly designate these malicious entities as their next-hop nodes for packet forwarding operations. When the subsequent data transmission phase commences, the true nature of the attack manifests—the malicious nodes intentionally discard all received data packets rather than forwarding them toward their intended destinations [5].

Malware Attack

Vehicular networks face significant security vulnerabilities during software maintenance operations. When onboard units (OBUs) and roadside units (RSUs) undergo software updates or receive patches, these maintenance channels create potential vectors for malware infiltration. Malicious software, including various forms of computer viruses, can be introduced during these update processes, subsequently disrupting normal network operations [46].

A distinguishing characteristic of malware attacks in vehicular networks is their typical origin. Unlike many other attack vectors, malware deployment is predominantly executed by malicious insiders rather than external threat actors. These insiders possess privileged knowledge of system architecture and update procedures, allowing them to craft targeted malicious payloads that can effectively compromise vehicular network components while evading detection mechanisms.

Jamming Attack

In vehicular networks, individual vehicles function as mobile nodes that exchange information through radio frequency (RF) signal transmission. This communication infrastructure, however, exhibits inherent vulnerabilities stemming from two fundamental characteristics: the intrinsic reliability limitations of mobile computing environments and the high scalability requirements of wireless systems.

Adversaries can exploit these vulnerabilities by deploying high-power interference signals directed at active communication channels. These deliberately generated interference patterns disrupt the normal signal reception capabilities of legitimate nodes within the network. The resulting signal degradation can significantly impair a node's ability to receive data packets, ranging from partial reduction in reception quality to complete communication failure [41].

Broadcast Tampering Attack

In false information injection attacks, adversaries compromise vehicular network integrity by deliberately introducing fraudulent security information or manipulating broadcast security messages within the system. This deceptive activity creates a distorted information environment that influences the decision-making processes of legitimate vehicles.

When exposed to such falsified data, affected vehicles make suboptimal or potentially dangerous choices based on what appears to be authentic information. The consequences of these manipulated decisions can manifest in various adverse outcomes, including elevated risk of traffic collisions or artificial traffic congestion along targeted roadways [47].

11.6.3 Data Integrity Attacks

Masquerading Attack

In this type of attack, malicious actors employ falsified identities to gain unauthorized access to the network. Once inside, they can modify or delete data packets being transmitted throughout the VANET. A representative example involves a malicious node impersonating an emergency vehicle, causing other vehicles to reduce speed or come to a complete stop [34, 10].

Replay Attack

In a replay attack, adversaries store legitimate V2V communications from a previous timeframe and rebroadcast them later when they are no longer valid. For instance, a malicious actor might capture messages about a traffic incident that occurred hours or days ago, then retransmit these messages to other vehicles in the network after the incident has been resolved. Vehicles receiving these replayed messages would interpret them as current information about road conditions, potentially altering their routes unnecessarily. This deception can artificially create traffic congestion in areas that would otherwise flow smoothly, as vehicles collectively respond to falsified data about road conditions that no longer exist [1].

Illusion Attack

In an illusion attack, malicious actors manipulate sensor readings on their own vehicle to generate fraudulent traffic information. They then broadcast these deceptive messages to nearby vehicles and infrastructure. The goal of this attack is to create artificial traffic congestion by causing other vehicles to reroute based on false information, ultimately disrupting normal traffic flow patterns [2].

Message Alteration Attack

In a manipulation attack, the adversary compromises data integrity by modifying network packets—adding unauthorized information, removing critical data, or discarding packets entirely. This interference undermines the fundamental reliability of transmitted information [16].

11.6.4 Confidentiality Attacks

Traffic Analysis Attack

In this type of attack, malicious actors monitor and analyze the messages exchanged in V2X communication networks. They systematically collect valuable information—such as vehicle locations and travel routes—which they can then use to influence or manipulate the behavior of other vehicles in the network [44]. This form of attack directly compromises data confidentiality within VANETs.

Eavesdropping Attack

The broadcast nature of wireless communication in VANETs makes transmissions between vehicles vulnerable to interception by unauthorized parties. Eavesdropping attacks represent a common security threat that primarily targets confidentiality at the network layer. During such attacks, malicious actors intercept sensitive information—such as vehicle location data that could be used for tracking—for unauthorized purposes [36].

11.7 Hands-on Exercises

In the section, we will first discuss in detail some real-world vulnerabilities of ROS2 with corresponding attacks. Then you will be able to implement this attacks yourself with step-by-by instructions. In this workshop, you need to have Python and both ROS and ROS2 installed.

11.7.1 Breaking the Access Control

Task Explore and exploit a vulnerability in ROS 2's access control mechanism by performing a privilege escalation attack using SROS2. You will simulate an unauthorized node attempting to publish to a protected topic.

Objective Set up a secure ROS 2 system using the SROS2 security plugins. Create a simple publisher-subscriber architecture with enforced access control. Then, configure a restricted attacker node that subscribes to `/attack_cmd` but is blocked from publishing. Finally, attempt a privilege escalation exploit by modifying the context to bypass ROS 2's security policy and enable unauthorized publication.

Deliverables

- A functioning SROS2-secured ROS 2 system with publisher and subscriber nodes.
- A restricted malicious node enclave configuration and demonstration of access denial.
- A documented method and demonstration of privilege escalation, including:
  - Modified policy or context structure.
  - Evidence of unauthorized message publication.
- A brief explanation of the exploit and its implications for ROS 2 security.

Recommended Papers to Read

1. Gao, C., Wang, G., Shi, W., Wang, Z., & Chen, Y. (2021). Autonomous driving security: State of the art and challenges. *IEEE Internet of Things Journal, 9*(10), 7572–7595. [18]
2. Deng, Y., Zhang, T., Lou, G., Zheng, X., Jin, J., & Han, Q. L. (2021). Deep learning-based autonomous driving systems: A survey of attacks and defenses. *IEEE Transactions on Industrial Informatics, 17*(12), 7897–7912. [11]
3. Sun, X., Yu, F. R., & Zhang, P. (2021). A survey on cyber-security of connected and autonomous vehicles (CAVs). *IEEE Transactions on Intelligent Transportation Systems, 23*(7), 6240–6259. [49]

References

1. Ali, I., Hassan, A., & Li, F. (2019). Authentication and privacy schemes for vehicular ad hoc networks (VANETs): A survey. *Vehicular Communications, 16*, 45–61.
2. Al-Kahtani, M. S. (2012). Survey on security attacks in vehicular ad hoc networks (VANETs). In *2012 6th international conference on signal processing and communication systems* (pp. 1–9). IEEE.
3. An, Z. et al. (2016). Development of mobile robot SLAM based on ROS. *International Journal of Mechanical Engineering and Robotics Research, 5*(1), 47–51.
4. Będkowski, J., et al. (2015). Open source robotic 3D mapping framework with ROS—robot operating system, PCL—point cloud library and cloud compare. In *2015 International conference on electrical engineering and informatics (ICEEI)* (pp. 644–649). IEEE.
5. Bibhu, V., et al. (2012). Performance analysis of black hole attack in VANET. *International Journal of Computer Network and Information Security, 4*(11), 47–54.
6. Cao, Y., et al. (2021). Invisible for both camera and lidar: Security of multi-sensor fusion based perception in autonomous driving under physical-world attacks. In *2021 IEEE symposium on security and privacy (SP)* (pp. 176–194). IEEE.
7. Checkoway, S., et al. (2011). Comprehensive experimental analyses of automotive attack surfaces. In *20th USENIX security symposium (USENIX Security 11).*
8. Chen, E.-C., et al. (2024). Overload: Latency attacks on object detection for edge devices. In *Proceedings of the IEEE/CVF conference on computer vision and pattern recognition* (pp. 24716–24725).
9. Corrigan, S. (2002). Introduction to the controller area network (CAN). In *Application report SLOA101* (pp. 1–17).
10. De Fuentes, J. M., González-Tablas, A. I., & Ribagorda, A. (2011). Overview of security issues in vehicular ad-hoc networks. In *Handbook of research on mobility and computing: Evolving technologies and ubiquitous impacts* (pp. 894–911). IGI Global.
11. Deng, Y., et al. (2021). Deep learning-based autonomous driving systems: A survey of attacks and defenses. *IEEE Transactions on Industrial Informatics, 17*(12), 7897–7912.
12. Di Luoffo, V., Michalson, W. R., & Sunar, B. (2018). Robot operating system 2: The need for a holistic security approach to robotic architectures. *International Journal of Advanced Robotic Systems, 15*(3), 1729881418770011 (2018)
13. Dimitriou, T., et al. (2016). Imposter detection for replication attacks in mobile sensor networks. *Computer Networks, 108*, 210–222.
14. Di Palma, C., et al. (2021). Security of camera-based perception for autonomous driving under adversarial attack. In *2021 IEEE security and privacy workshops (SPW)* (pp. 243–243). IEEE.

15. Eryigit, C., & Uyar, S. (2008). Integrating agents into data-centric naval combat management systems. In *2008 23rd international symposium on computer and information sciences* (pp. 1–4). IEEE.
16. Gadkari, M. Y., & Sambre, N. B. (2012). VANET: Routing protocols, security issues and simulation tools. *IOSR Journal of Computer Engineering, 3*(3), 28–38.
17. Gao, Y., et al. (2019). A distributed network intrusion detection system for distributed denial of service attacks in vehicular ad hoc network. *IEEE Access, 7*, 154560–154571.
18. Gao, C. et al. (2021). Autonomous driving security: State of the art and challenges. *IEEE Internet of Things Journal, 9*(10), 7572–7595.
19. Guette, G., & Ducourthial, B. (2007). On the Sybil attack detection in VANET. In *2007 IEEE international conference on Mobile Adhoc and sensor systems* (pp. 1–6). IEEE.
20. Hannan, M. A., Hussain, A., & Samad, S. A. (2010). System interface for an integrated intelligent safety system (ISS) for vehicle applications. *Sensors, 10*(2), 1141–1153.
21. Hasrouny, H., et al. (2017). VANet security challenges and solutions: A survey. In *Vehicular Communications, 7*, 7–20.
22. Herberth, R., et al. (2019). Automated scheduling for optimal parallelization to reduce the duration of vehicle software updates. *IEEE Transactions on Vehicular Technology, 68*(3), 2921–2933.
23. Howard, M. (2003). *Fending off future attacks by reducing attack surface*. Microsoft MSDN.
24. Ioannides, R. T., Pany, T., Gibbons, G. (2016). Known vulnerabilities of global navigation satellite systems, status, and potential mitigation techniques. *Proceedings of the IEEE, 104*(6), 1174–1194.
25. Jeong, S.-Y., et al. (2017). A study on ros vulnerabilities and countermeasure. In *Proceedings of the companion of the 2017 ACM/IEEE international conference on human-robot interaction* (pp. 147–148).
26. Kim, J. et al. (2018). Security and performance considerations in ros 2: A balancing act. Preprint, arXiv:1809.09566 .
27. Koscher, K. et al. (2010). Experimental security analysis of a modern automobile. In *2010 IEEE symposium on security and privacy* (pp. 447–462). IEEE.
28. Lawrence, J. (2020). ROS2 prevalance and security. In *Report CSEC 793*. Rochester Institute of Technology, Rochester, NY, USA
29. Lim, B. S., Keoh, S. L., & Thing, V. L. L. (2018). Autonomous vehicle ultrasonic sensor vulnerability and impact assessment. In *2018 IEEE 4th world forum on internet of things (WF-IoT)* (pp. 231–236). IEEE.
30. Liu, J., Yan, C., & Xu, W. (2016). Can you trust autonomous vehicles: Contactless attacks against sensors of self-driving vehicles. In *Las Vegas: DEF CON, 24*.
31. Liu, S., et al. (2018) *Creating autonomous vehicle systems*. Springer.
32. Liu, S., et al. (2019). Edge computing for autonomous driving: Opportunities and challenges. *Proceedings of the IEEE, 107*(8), 1697–1716.
33. McClean, J., et al. (2013). A preliminary cyber-physical security assessment of the robot operating system (ros). In *Unmanned systems technology xv* (Vol. 8741, pp. 341–348). SPIE.
34. Mejri, M. N., Ben-Othman, J., & Hamdi, M. (2014). Survey on VANET security challenges and possible cryptographic solutions. *Vehicular Communications, 1*(2), 53–66.
35. Miller, C., & Valasek, C. (2015). Remote exploitation of an unaltered passenger vehicle. *Black Hat USA, 2015*(S 91), 1–91.
36. Mokhtar, B., & Azab, M. (2105). Survey on security issues in vehicular ad hoc networks. *Alexandria Engineering Journal, 54*(4), 1115–1126 (2015)
37. Ning, J., et al. (2019). Attacker identification and intrusion detection for in-vehicle networks. *IEEE Communications Letters, 23*(11), 1927–1930 (2019)
38. Pardo-Castellote, G. (2003). Omg data-distribution service: Architectural overview. In *23rd international conference on distributed computing systems workshops, 2003. Proceedings* (pp. 200–206). IEEE.
39. Petit, J., & Shladover, S. E. (2014). Potential cyberattacks on automated vehicles. *IEEE Transactions on Intelligent transportation systems, 16*(2), 546–556 (2014)

40. Petit, J., et al. (2015). Remote attacks on automated vehicles sensors: Experiments on camera and lidar. *Black Hat Europe, 11*, 995.
41. Rawat, R., & Sharma, D. (2015). Impact of jamming attack in vehicular ad hoc networks. *International Journal of Advanced Research in Computer and Communication Engineering, 4*(4), 457–461.
42. Rawat, A., Sharma, S., & Sushil, R.. (2012). VANET: Security attacks and its possible solutions. *Journal of Information and Operations Management, 3*(1), 301–304.
43. Sakiz, F., & Sen, S. (2017). A survey of attacks and detection mechanisms on intelligent transportation systems: VANETs and IoV. *Ad Hoc Networks, 61*, 33–50.
44. Sheikh, M. S., & Liang, J. (2019). A comprehensive survey on VANET security services in traffic management system. *Wireless Communications and Mobile Computing, 2019*(1), 2423915.
45. Shin, H., et al. (2017). Illusion and dazzle: Adversarial optical channel exploits against lidars for automotive applications. In *Cryptographic hardware and embedded systems–CHES 2017: 19th international conference, Taipei, Taiwan, September 25–28, 2017, Proceedings* (pp. 445–467). Springer.
46. Shukla, D., et al. (2016). Security and attack analysis for vehicular ad hoc network—A survey. In *2016 international conference on computing, communication and automation (ICCCA)* (pp. 625–630.). IEEE.
47. Sumra, I. A., et al. (2013). Classification of attacks in vehicular ad hoc network (vanet). *International Information Institute (Tokyo) Information, 16*(5), 2995.
48. Sumra, I. A., Hasbullah, H. B., & AbManan, J.-L. B. (2014). Attacks on security goals (confidentiality, integrity, availability) in VANET: a survey. In *Vehicular ad-hoc networks for smart cities: First international workshop, 2014* (pp. 51–61.). Springer.
49. Sun, X., Yu, F. R., & Zhang, P. (2021). A survey on cyber-security of connected and autonomous vehicles (CAVs). *IEEE Transactions on Intelligent Transportation Systems, 23*(7), 6240–6259.
50. Thing, V. L. L., & Wu, J. (2016). Autonomous vehicle security: A taxonomy of attacks and defences. In *2016 IEEE international conference on internet of things (ithings) and IEEE green computing and communications (greencom) and IEEE cyber, physical and social computing (cpscom) and IEEE smart data (smartdata)* (pp. 164–170.). IEEE.
51. Wang, B. et al. (2017). *Driver identification using vehicle telematics data*. Technical Report, SAE Technical Paper.
52. White, R., Christensen, H. I., & Quigley, M. (2016). SROS: Securing ROS over the wire, in the graph, and through the kernel. Preprint, arXiv:1611.07060.
53. Xu, W., et al. (2018). Analyzing and enhancing the security of ultrasonic sensors for autonomous vehicles. *IEEE Internet of Things Journal, 5*(6), 5015–5029.
54. Young, C., et al. (2019). Survey of automotive controller area network intrusion detection systems. *IEEE Design & Test, 36*(6), 48–55.
55. Yousuf, A., et al. (2015). Introducing kinematics with robot operating system (ROS). In *2015 ASEE annual conference & exposition* (pp. 26–1024).
56. Zeadally, S., et al. (2012). Vehicular ad hoc networks (VANETS): Status, results, and challenges. *Telecommunication Systems, 50*, 217–241.

Chapter 12
Simulation and Testing Techniques

12.1 Introduction

This chapter delves into the foundational role of simulation and testing techniques in the development and deployment of autonomous vehicle (AV) systems. As AVs must operate in a vast range of dynamic, unpredictable environments, simulation emerges as a crucial method for exploring and refining system behaviors before physical implementation. It provides a virtual proving ground where engineers can rigorously test and optimize the performance of perception, decision-making, and control systems under a wide variety of conditions—including scenarios that would be hazardous, cost-prohibitive, or nearly impossible to stage in the real world. From sudden pedestrian crossings to sensor malfunctions in adverse weather, simulation makes it feasible to test the full spectrum of operational situations an AV may encounter. When coupled with formal verification techniques and software validation protocols, simulation becomes a powerful enabler of safe, scalable, and continuous AV development, dramatically accelerating innovation while minimizing risk.

Simulation serves as a foundation for AV development by allowing developers to test perception, planning, and control algorithms in a controlled and repeatable setting. It significantly reduces cost, risk, and development time, and it supports testing in scenarios ranging from normal traffic conditions to rare edge cases such as sudden pedestrian crossings or sensor failures.

Key advantages include:

- **Risk-free evaluation of dangerous situations:** Simulation enables the recreation of hazardous driving conditions—such as sudden obstacles, high-speed evasive maneuvers, or sensor malfunctions—without endangering human lives or property. This allows teams to refine algorithms under high-risk circumstances that would be too dangerous to trial in real environments. For example, researchers can simulate a vehicle's response to a child running into the street

W. Shi, Y. He, *Introduction to Autonomous Driving*,
https://doi.org/10.1007/978-3-031-99485-2_12

from behind a parked car, a situation nearly impossible to stage safely in real life.

- **Scalability for testing thousands of scenarios:** Developers can rapidly simulate a wide array of driving scenarios, ranging from common intersections to complex urban networks, and test how AVs respond to different traffic densities, road geometries, and actor behaviors. This massive scalability allows for statistical confidence in safety performance. For instance, a company might simulate 10,000 variations of a merging scenario on a highway to identify patterns of failure and optimize merging logic accordingly.
- **Repeatability for consistent debugging:** Identical simulations can be run repeatedly to analyze specific behaviors, reproduce bugs, or evaluate the effect of incremental software changes. This level of consistency is difficult to achieve in physical testing, where environmental and situational variables often change between trials. As an example, a developer might rerun a sensor fusion test involving fog and night-time lighting to ensure that perception bugs introduced by a new LiDAR firmware update are effectively resolved.
- **Flexibility for changing environmental parameters:** Simulation allows developers to easily manipulate environmental factors such as weather, time of day, and road surface conditions. This makes it possible to evaluate system robustness across a spectrum of operating conditions and tailor testing to regional or seasonal needs. For example, AV systems intended for snowy climates can be evaluated in white-out conditions or on icy roads to measure braking accuracy and control stability. Simulation allows developers to easily manipulate environmental factors such as weather, time of day, and road surface conditions. This makes it possible to evaluate system robustness across a spectrum of operating conditions and tailor testing to regional or seasonal needs.

Simulation plays a pivotal role across the entire AV development lifecycle, serving as a continuous thread from conception to deployment. During the initial design phase, simulation provides a flexible and dynamic testing ground to explore algorithm prototypes in both idealized conditions and worst-case scenarios. Developers can test different sensor placement strategies to assess field-of-view coverage and minimize blind spots, analyze the effects of various vehicle control schemes, and simulate environmental factors such as traffic density, lighting changes, and road topologies before any physical components are built. For example, before selecting a LiDAR configuration, engineers can evaluate detection accuracy across multiple mounting positions and in varied driving environments. This early-stage simulation not only mitigates design flaws before they manifest physically but also accelerates the innovation cycle by enabling rapid hypothesis testing and iteration without material costs or safety risks. Through iterative feedback loops in the simulated environment, teams can refine architectural choices and system behaviors well before entering more costly and constrained stages of prototyping.

As development progresses into the prototyping phase, simulation becomes essential for robust verification of software modules using synthetic datasets that replicate a wide array of operational conditions. These datasets can be tailored to stress-test specific capabilities, such as obstacle detection in heavy fog, lane marking identification at twilight, or pedestrian tracking in crowded crosswalks. Perception algorithms are rigorously assessed against varying lighting, weather, and occlusion scenarios to ensure they maintain detection accuracy and classification reliability across contexts. Likewise, planning algorithms are examined for their consistency and safety in decision-making during complex maneuvers like roundabout navigation, dynamic merging on highways, or four-way stop interactions with other vehicles.

Moreover, simulations during this phase allow for the emulation of rare edge cases and fault injection testing—such as sensor blackouts or conflicting GPS data—to examine how resilient each module is under failure conditions. Controlled virtual environments ensure that each software component performs to specification not just in isolation, but in dynamic and sometimes adversarial conditions. By identifying unexpected behaviors early in the development process, developers minimize costly debugging and rework during real-world trials. This level of pre-integration validation also helps establish confidence in the system's readiness for more complex, hardware-in-the-loop and on-road evaluations.

In the integration phase, simulation allows for comprehensive testing of how multiple AV subsystems—such as localization, sensor fusion, trajectory planning, and control—interact and function as a unified system. At this stage, it becomes essential to detect hidden interdependencies that may not be visible when modules are tested in isolation. For instance, a slight timing mismatch between sensor fusion and control commands could cause trajectory deviations that only emerge under real-time constraints. Simulating these integrated systems under variable traffic conditions, terrain complexity, and communication delays can uncover edge-case failures or subtle conflicts between algorithmic priorities.

Integration testing also helps assess whether updates to one subsystem inadvertently degrade the performance of another. For example, an update to the perception module might introduce additional latency or jitter in the data stream, which could in turn impair the responsiveness and accuracy of the planning module. This can result in delays in trajectory generation or incorrect path decisions, particularly in fast-evolving scenarios like merging or evasive maneuvers. These kinds of cascading effects, often rooted in complex timing dependencies or software interactions, are difficult to detect through unit tests or isolated simulations alone. System-level simulation provides a synchronized and holistic environment in which these subtle yet critical issues can be observed, diagnosed, and corrected before deployment. By replicating real-time interactions among modules and simulating environmental pressures such as traffic density and communication lag, engineers gain the insights needed to ensure system coherence and resilience under stress.

Following integration, simulation becomes a powerful and indispensable tool for regression testing. Developers can systematically rerun a library of previously validated scenarios—such as lane merges, obstacle avoidance, or complex intersection navigation—to confirm that new software updates or hardware modifications do not reintroduce previously resolved errors. This continuous validation process is vital for identifying unintended side effects, especially when modifications to one subsystem may have ripple effects on others.

In safety-critical applications such as autonomous driving, even a subtle change in perception accuracy or trajectory planning latency can compromise system integrity. Regression testing in simulation ensures that the system's baseline performance remains consistent across iterations, maintaining confidence in safety guarantees and long-term reliability. For instance, when updating the object detection module to improve nighttime visibility, regression tests can immediately reveal whether detection under foggy or rainy conditions has been inadvertently degraded.

Moreover, regression testing supports traceability and version control, enabling teams to track the evolution of system behavior over time. This historical analysis is invaluable for audits, certifications, and root-cause investigations when unexpected performance issues emerge later in the deployment lifecycle.

Simulation also supports the creation of digital twins—high-fidelity, dynamic virtual replicas of real-world vehicles and their operating environments. These digital twins are not static models but are continuously updated with real-time data from the physical counterparts, enabling synchronized testing and predictive diagnostics. By simulating a vehicle's entire hardware and software stack within a digital twin, developers can conduct comprehensive testing of interactions between mechanical systems, embedded software, and sensor inputs under a wide range of virtual conditions.

Digital twins enable real-time monitoring and system performance forecasting, which is especially valuable for identifying anomalies before they manifest physically. For example, by modeling battery temperature behavior and correlating it with energy consumption patterns in the virtual twin, engineers can preemptively address thermal management issues. Similarly, digital twins allow for fine-tuning of control strategies based on localized traffic conditions or wear and tear trends. This integration of virtual and real data creates a powerful feedback loop that enhances system robustness, streamlines maintenance, and accelerates continuous improvement cycles in both development and post-deployment phases.

Advanced simulation tools now incorporate synthetic sensor data generation, photorealistic rendering, and reinforcement learning environments, revolutionizing both the validation and training processes for autonomous systems. Synthetic data—artificially generated from simulation rather than real-world sensors—enables the creation of diverse and highly controlled datasets that reflect challenging or underrepresented driving conditions, such as unusual pedestrian behaviors, rare vehicle types, or extreme weather events. These datasets help bridge the gap where real-world data is scarce, expensive, or difficult to obtain ethically.

Photorealistic rendering engines enhance the realism of simulation environments, allowing computer vision algorithms to be trained and validated in lifelike urban and rural scenes. Shadows, reflections, varying light angles, and atmospheric effects can be precisely tuned to stress-test visual recognition systems.

Additionally, reinforcement learning environments integrated into simulators allow autonomous agents to learn behaviors through repeated interaction with dynamic virtual settings. These environments support policy learning for tasks like adaptive cruise control, evasive maneuvers, or cooperative merging, with the benefit of safe exploration and accelerated convergence.

Collectively, these capabilities elevate simulation from a passive testing environment to an active platform for intelligent system training and behavioral optimization across the AV development pipeline.

Collaboration across disciplines is enhanced through simulation as well. Engineers, data scientists, safety analysts, and regulators can interact with the same simulated environments to understand system behaviors, debug performance anomalies, or review safety case evidence.

As the industry moves toward regulatory oversight of AV safety, simulation also provides the basis for formal documentation and compliance. Standards such as ISO 26262 and UL 4600 increasingly rely on simulation evidence to support functional safety claims. In particular, scenario coverage analysis, traceability, and reproducibility enabled by simulation are integral to building the argument for AV safety assurance.

12.2 Types of Simulation

Simulation in the context of autonomous vehicle development is not monolithic; it encompasses a spectrum of methodologies designed to test, validate, and train various layers of the system architecture. These types differ in terms of their complexity, fidelity, and integration level with real hardware and software. Understanding these categories—Model-in-the-Loop (MIL), Software-in-the-Loop (SIL), and Hardware-in-the-Loop (HIL)—is essential for constructing a comprehensive and scalable testing pipeline.

Each type of simulation builds upon the other, progressively incorporating more elements of the final system until full integration is achieved. This layered approach allows developers to debug, optimize, and verify AV systems at every level of abstraction, from algorithmic logic to hardware execution. In the sections that follow, we will explore each simulation type in detail, including their objectives, implementation, and roles within the AV development lifecycle.

12.2.1 Model-in-the-Loop (MIL)

Model-in-the-Loop (MIL) simulation represents the foundational stage of autonomous vehicle (AV) system testing. It is critical because it enables early validation of functional correctness before committing to complex software integration or hardware deployment. Engineers use MIL to construct high-level representations of algorithms and system components, which can then be tested in a mathematically controlled environment. This approach supports iterative development, risk mitigation, and design space exploration from the outset of

Table 12.1 Comparison of Model-in-the-Loop (MIL) simulation platforms

Platform	Type	Key strengths	Integration features	Ideal use case
MATLAB Simulink	Proprietary	Visual modeling, block diagrams, continuous/discrete dynamics, code generation	Simulink toolchain, real-time visualization, FMI	Control design, rapid prototyping, algorithm tuning
SCANeR Studio	Proprietary	High-fidelity traffic simulation, environmental realism, scenario diversity	ASAM OpenSCENARIO, vehicle dynamics, co-simulation	Early behavioral validation, safety concept design
Modelica (OpenModelica, Dymola)	Open-source proprietary	Multi-domain system modeling, acausal connections, physical modeling	FMI compliance, customizable libraries	Energy systems, thermal modeling, AV subsystems
SimPy Python-based MIL	Open-source	Lightweight, scriptable event-based simulation, integration with analytics libraries	Python integration, fast iteration	Custom simulators for specific algorithm testing

the AV development lifecycle. MIL provides a cost-effective and computationally efficient strategy for identifying flaws and performance limitations at the earliest possible stage (Table 12.1).

Tools such as MATLAB/Simulink [16] are widely used in MIL due to their ability to model hybrid systems involving both continuous and discrete dynamics. These tools employ graphical block-diagram representations, which help engineers build hierarchical systems, model time-series sensor data, and simulate closed-loop control architectures. One of the most important features of Simulink is its code generation capability, which facilitates a smoother transition from abstract models to executable code. Toolboxes for areas like sensor fusion and physical modeling further extend its relevance to autonomous driving systems. These environments enable developers to replicate typical AV operations—such as lane keeping, obstacle avoidance, and adaptive cruise control—while tuning parameters under simulated conditions.

The primary advantage of MIL lies in its ability to decouple logic validation from the constraints of physical hardware. Developers can run thousands of test cases across a range of conditions, performing 'what-if' analyses and sensitivity studies that would be infeasible or unsafe in real-world testing. MIL is especially powerful for validating control algorithms, perception pipelines, and decision-making logics in isolation. However, this decoupling introduces inherent limitations: the abstract nature of models means MIL lacks fidelity in capturing real-time execution characteristics, memory management, and hardware-induced latency. Thus, while it promotes clarity, reproducibility, and speed, MIL results must eventually be validated under more realistic conditions.

A prototypical application of MIL might involve simulating a lane-keeping assist system. By constructing a model of the road environment, vehicle kinematics, and sensor feedback, engineers can evaluate how control logic responds to lateral drift or curvature changes. In this setting, MIL supports debugging and performance evaluation without requiring real sensor data or an instrumented test vehicle. Formal verification methods, along with stochastic techniques such as Monte Carlo simulation, allow the exploration of rare edge cases, contributing to more robust downstream integration.

Modern MIL environments, such as SCANeR Studio [19], extend the traditional boundaries of abstraction by incorporating high-fidelity environmental elements. These platforms simulate complex traffic dynamics, varying weather conditions, and heterogeneous agents, all within the MIL framework. The result is a hybrid form of simulation that preserves MIL's speed and clarity while approximating some of the realism typically reserved for Hardware-in-the-Loop (HIL) or field tests. This increased fidelity during early development stages reduces design iteration time and improves the reliability of later testing stages.

The core strength of MIL lies in its ability to provide deterministic and repeatable simulations. By abstracting away real-world variability, MIL enables precise testing of specific logic flows and system behaviors. Subsystems like motion planning, trajectory optimization, or vehicle-pedestrian interaction modules can be tested independently, promoting modular design and validation. Nevertheless, this abstraction is also its weakness: MIL does not account for real-time scheduling conflicts, sensor noise, or the asynchronous nature of physical-world interactions. Consequently, MIL cannot replace higher-fidelity simulations or real-world testing but must be viewed as a foundational component in a multi-stage validation pipeline.

Recent developments in co-simulation frameworks have made MIL even more powerful. These approaches enable different domain models—such as vehicle dynamics, sensor emulation, and traffic simulation—to run concurrently while exchanging data in real-time. This interoperability allows developers to observe emergent system behavior that arises from cross-domain interactions, such as complex merging scenarios or cooperative lane changes. Such capabilities are critical for developing AV features that depend on distributed intelligence or V2X communication.

In parallel, standardization efforts like ASAM OpenSCENARIO® have increased the portability and reusability of simulation scenarios. These standards ensure that MIL test cases can be easily shared, reused, or integrated into broader safety assessment workflows. As the field evolves, MIL is no longer confined to early-stage design but increasingly serves as a tool for continuous integration and safety assurance. Engineers now leverage MIL not only for prototyping, but also for proactive detection of system weaknesses and verification against industry safety metrics.

12.2.2 Software-in-the-Loop

Software-in-the-Loop (SIL) simulation is the next stage of fidelity, where the actual production software—such as perception, localization, and planning modules—is tested within a simulated environment. Unlike Model-in-the-Loop (MIL), SIL involves compiling and executing the system's real code, often in the same programming language and architecture used for deployment [6]. This ensures behavioral consistency with real-world implementations and supports early detection of implementation-level bugs, integration mismatches, and performance bottlenecks. SIL bridges the gap between high-level algorithmic design and full deployment, enabling verification of real-time responses, threading behavior, and interface correctness in a safe, virtual setting.

SIL enables developers to validate how software modules interact internally and with simulated sensor inputs or vehicle dynamics. It is commonly implemented using modular software platforms such as ROS2, Docker, or native simulation environments that support message passing, modular system integration, and real-time visualization [11]. Simulators like CARLA [3], LGSVL [18], and Gazebo allow developers to generate synthetic sensor data—including camera, LiDAR, radar, and IMU—which mimics real-world signal patterns. These data streams are then processed by the actual AV software stack, facilitating end-to-end system evaluation in repeatable and configurable conditions.

SIL testing is essential for uncovering issues that remain invisible during MIL. It allows for rigorous stress testing of real software modules using thousands of systematically varied scenarios. For example, developers can explore how latency in perception pipelines affects planning stability, or how GPS spoofing impacts localization fidelity [1]. These tests expose vulnerabilities such as memory overflows, deadlocks, or improper sensor calibration that are difficult to capture in abstract modeling. By testing the compiled code in a deterministic simulation, SIL also supports performance profiling and resource usage analysis, which are crucial for embedded deployment.

Moreover, SIL supports continuous integration (CI) and automated regression testing by embedding simulation runs into the development pipeline. When integrated with CI tools, each software commit can be automatically validated against a predefined suite of scenarios, ensuring that code changes do not introduce regressions. This makes SIL a critical enabler of DevOps workflows in AV development [5]. It supports nightly builds, branch testing, and staged deployment strategies, thus shortening iteration cycles and improving code quality (Table 12.2).

Recent advancements have significantly expanded the capabilities of SIL platforms. Cloud-based solutions like dSPACE VEOS [5] now support distributed simulation of virtual electronic control units (vECUs) across scalable infrastructure. VEOS excels at hardware abstraction, enabling large-scale system testing without dedicated hardware. AVL SIL [2] further supports closed-loop simulations of multi-domain vehicle systems, allowing engineers to validate powertrain interactions, control loops, and fault responses with high fidelity.

Table 12.2 Comparison of Software-in-the-Loop (SIL) simulation platforms

Platform	Type	Key strengths	Integration features	Ideal use case
CARLA [3]	Open-source	Realistic 3D urban environments, photorealistic rendering, ROS compatibility	Python API, scenario engine, ROS bridge	Academic research, perception and planning testing
SVL Simulator [18]	Open-source	High-fidelity traffic scenarios, supports Apollo and Autoware	Real-time sensor streaming, scenario editing tools	Full-stack AV software validation
Gazebo [11]	Open-source	Modular robotics simulation, strong ROS integration	Plugin-based architecture, physics simulation	Localization, SLAM, and robotics control research
SUMO [12]	Open-source	Microscopic traffic simulation, realistic signal and vehicle behavior	Co-simulation with CARLA, VISTA, OpenCDA, and TraCI API	Traffic flow modeling, urban planning, AV scenario realism
OpenCDA [24]	Open-source	Cooperative driving automation framework with scenario sharing and coordination logic	CARLA/VISTA integration, ROS2, SUMO co-simulation	V2X research, cooperative maneuvers, policy evaluation
dSPACE VEOS [5]	Proprietary	Scalable cloud simulation, virtual ECU modeling	CI/CD pipelines, ASAM compliance, hardware abstraction	Industrial regression testing, integration validation
Siemens Prescan [21]	Proprietary	Physics-based sensor and V2X modeling	Simulink integration, Open-SCENARIO, traffic simulation	ADAS/AV safety verification, compliance testing
AVL SIL suite [2]	Proprietary	Multi-domain vehicle simulation, fault injection	Closed-loop co-simulation, regression automation	Powertrain, thermal systems, and ADAS validation
ETAS COSYM [6]	Proprietary	Real-time embedded system testing, cloud scalability	DevOps toolchain, FMI/XIL compatibility	Integrated embedded systems and ECU orchestration

Siemens Prescan [21] stands out for its physics-based sensor modeling and rich environmental simulation. It supports comprehensive V2X communication modeling and scenario creation, offering built-in support for ADAS and AV safety validation workflows. Prescan integrates with MATLAB/Simulink and supports standard scenario formats like ASAM OpenSCENARIO, making it a powerful choice for verification aligned with regulatory standards.

ETAS COSYM [6] offers real-time co-simulation capabilities and is tailored for embedded systems testing in DevOps contexts. It provides full support for FMI and ASAM XIL standards, ensuring interoperability with a wide range of model-based and code-based testing tools. COSYM excels in systems-level validation where multiple software components and ECUs must operate in sync.

Open-source tools also play a vital role. CARLA [3] is an Unreal Engine-based simulator that supports photorealistic rendering, sensor simulation, and full-stack AV integration. It is particularly strong in perception and planning research, thanks to its Python API and support for customizable maps and actor behaviors.

LGSVL [18] offers modular integration with Apollo and Autoware stacks, emphasizing high-fidelity traffic modeling and sensor realism. It enables closed-loop testing for entire AV pipelines and supports ROS bridges, allowing easy insertion into academic or industrial development pipelines.

Gazebo [11], often used with ROS2, provides a modular and plugin-driven architecture. It is especially useful for research in localization, mapping, and robotic control systems. Although less visually refined than CARLA, Gazebo's flexibility and strong community support make it an ideal platform for rapid experimentation and prototyping.

Traffic-level simulation can be enriched using SUMO [12], an open-source tool that models realistic traffic flow and signal control, and integrates with AV simulators through co-simulation frameworks

Frameworks like OpenCDA [24] integrate co-simulation with cooperative driving automation, enabling rigorous scenario testing for V2V and V2I communication, and extending SIL capabilities to include multi-agent coordination and connected infrastructure.

Together, these platforms demonstrate the rich landscape of SIL simulation environments. Open-source platforms such as CARLA, SVL Simulator, and Gazebo provide flexible, community-driven environments suitable for academic research and prototyping. They are easily customizable and integrate with open middleware, but may lack some of the standard compliance or industrial robustness needed for safety-critical applications.

Conversely, proprietary platforms like dSPACE VEOS, Siemens Prescan, AVL SIL Suite, and ETAS COSYM offer comprehensive toolchains, standards compliance, and traceability features required for commercial AV development and certification. They support scalable simulations, co-simulation frameworks, and test automation, aligning with ISO 26262 and other regulatory standards.

The selection between these tools ultimately depends on project scale, budget, compliance needs, and the maturity of the development process. In practice, hybrid toolchains combining both open-source and commercial platforms are increasingly used to balance innovation and reliability in AV system validation.

12.2.3 Hardware-in-the-Loop (HIL)

Hardware-in-the-Loop (HIL) simulation introduces real physical components—such as ECUs, sensors, or actuators—into the testing loop. This stage serves as a high-fidelity bridge between software validation and full vehicle testing, combining real-time hardware feedback with simulated environments. While Software-in-the-Loop (SIL) simulations allow for code-level validation within virtual testbeds, HIL takes the next step by integrating actual hardware, capturing electrical, physical, and temporal nuances that software-only setups cannot. This transition substantially improves test coverage by enabling verification of real-world signal behaviors, interface protocols, and timing compliance under realistic operational loads. It bridges key gaps in simulation fidelity by reproducing conditions where delays, signal jitter, or actuator inconsistencies could compromise safety or functionality—scenarios that are difficult to observe in purely virtual environments (Table 12.3).

HIL simulation enables developers to evaluate not just code performance, but also the physical and electrical behavior of components under dynamic conditions. Unlike purely virtual simulations, HIL setups replicate the electrical and timing characteristics of in-vehicle systems, making them critical for testing latency-sensitive or safety-critical modules. Simulation tools in HIL enable these capabilities by generating synthetic sensor signals, injecting test scenarios, and synchronizing software execution with real-world timing constraints. These tools often support real-time operating systems, hardware abstraction layers, and closed-loop testing circuits to create realistic operational environments. As such, HIL is indispensable for evaluating integration fidelity, verifying hardware compliance with performance specifications, and ensuring robust system behavior under operational stress. These tools often support real-time operating systems, hardware abstraction layers, and closed-loop testing circuits to create realistic operational environments. As such, HIL is indispensable for evaluating integration fidelity, verifying hardware compliance with performance specifications, and ensuring robust system behavior under operational stress.

In HIL setups, simulated sensor signals are fed into real AV hardware systems, which respond as if they were operating in a real-world environment. For instance, a braking control algorithm running on a physical ECU can be triggered by a simulated emergency stop scenario, allowing engineers to measure braking latency, actuation accuracy, and system responsiveness. These simulations emulate real sensor inputs—such as wheel speed, acceleration, or proximity signals—using signal conditioning hardware and protocol converters that replicate the timing, format, and electrical characteristics of real-world data. This hardware-in-the-loop configuration allows developers to test system behavior under precise and reproducible conditions, observe system outputs via diagnostic interfaces, and capture performance metrics like actuation delay or communication lag. Additionally, engineers can inject faults such as sensor dropouts or erroneous values to assess how the system responds to degraded inputs, making HIL essential for verifying both functional integrity and failure-handling capabilities.

Table 12.3 Comparison of Hardware-in-the-Loop (HIL) simulation platforms

Platform	Type	Key strengths	Integration capabilities	Ideal use case
dSPACE SCALEXIO	Proprietary	Real-time simulation with high-speed I/O and sensor-realistic modeling	MATLAB/ Simulink, AUTOSAR, CAN, LIN, FlexRay	ECU testing, ADAS development, system-level validation
NI VeriStand	Proprietary	Real-time testing with strong LabVIEW integration and signal conditioning tools	LabVIEW, TestStand, NI PXI systems	Embedded control and sensor integration validation
Ansys AVxcelerate	Proprietary	Sensor physics modeling with high-fidelity environmental simulation	Integration with third-party platforms, real sensor emulation	Advanced sensor testing and scenario-based safety validation
IPG CarMaker	Proprietary	Closed-loop testing with real-time vehicle dynamics and road modeling	Simulink, FMI, CAN, ROS	Driving function validation, ECU integration, and vehicle dynamics assessment
AirSim (with hardware I/O)	Open-source	High-fidelity Unreal Engine-based simulation with Python/C++ APIs	Custom hardware bridging, ROS integration	AV and drone prototyping with physical I/O extensions
Project Chrono (HIL-capable)	Open-source	Multi-physics simulation and vehicle dynamics engine	ROS, FMI, Simulink integration	Robotics and vehicle dynamics research in HIL contexts

Real-time constraints are a defining feature of HIL simulation, making timing precision not only important but essential for meaningful test outcomes. In a HIL setup, simulated sensor signals must reach physical hardware—such as ECUs or actuators—at exactly the right intervals to reflect the temporal dynamics of real-world operation. Any lag or jitter in signal delivery can skew test results or fail to reveal critical issues. Therefore, communication protocols such as CAN, LIN, FlexRay, or automotive Ethernet must be implemented with nanosecond-level synchronization to reflect authentic bus behavior.

Simulation tools involved in HIL must generate and process these signals in real time, accounting for feedback from the physical hardware. For example, if an ECU outputs a braking command, the HIL platform must measure its response instantly and adjust environmental models accordingly, completing a realistic feedback loop. To support this, specialized hardware components like signal conditioners, digital-to-analog converters, power emulators, and diagnostic oscilloscopes are employed to ensure electrical characteristics match in-vehicle expectations. These tools emulate the vehicle's electrical network topology, power domains, and transient behaviors under load, making the virtual environment indistinguishable from a live automotive system. This fidelity is indispensable for verifying functional performance, safety compliance, and fault response mechanisms in mission-critical AV subsystems.

HIL is particularly vital for safety-critical systems—such as fail-safe braking, redundant power control, or V2X communications—where software behavior must align precisely with hardware limitations and real-world demands. These systems require deterministic timing, fault tolerance, and robust fallback mechanisms, all of which can be rigorously tested using HIL simulation. The ability to interface with physical hardware enables engineers to assess how real components respond to abnormal operating conditions and unexpected environmental inputs.

Fault injection in HIL—such as voltage sags, sensor outages, CAN message corruption, or actuator lag—permits controlled evaluation of system resilience and recovery protocols. These tests are designed to simulate plausible fault conditions and observe how the system detects, isolates, and responds to these anomalies. For example, a simulated CAN message dropout may trigger a fallback behavior in the motion planning module, while an actuator lag might engage an alert mechanism and reconfigure control setpoints. Common recovery strategies include redundant computation paths, watchdog timers, fault-tolerant state machines, and degraded-mode operation, all of which can be tested and validated under HIL. This testing uncovers vulnerabilities in both software logic and hardware interfaces that may not be evident in virtual environments alone. Moreover, the ability to induce and isolate faults enables root-cause analysis, informs the design of fail-operational architectures, and validates compliance with safety standards that require demonstrated fault response capabilities.

As such, HIL plays a central role in pre-certification testing, particularly in compliance with safety standards like ISO 26262, especially for achieving Automotive Safety Integrity Levels (ASILs) ranging from A to D, where D represents the highest degree of hazard and risk. HIL simulations enable rigorous testing of diagnostic coverage, fault detection, fail-silent behavior, and real-time response performance—critical elements for satisfying ISO 26262 functional safety requirements. Similarly, HIL supports validation against AUTOSAR compliance for software and hardware component interoperability and modularity. It is also essential for final hardware verification, production-level validation, and end-of-line testing, where consistent, repeatable evaluations are needed to confirm that systems meet design specifications and regulatory safety criteria before vehicle integration or release.

Below is a list of key HIL simulation platforms with their strengths and specific uses:

- dSPACE SCALEXIO: A leading commercial HIL system that combines real-time execution capabilities with high-speed analog and digital I/O, making it suitable for the most demanding AV and ADAS applications. SCALEXIO

supports advanced sensor emulation for devices like cameras, radar, and LiDAR through signal generation modules and programmable interfaces. It integrates tightly with MATLAB/Simulink, allowing developers to deploy models directly onto the real-time system and interface with existing toolchains. Communication protocols like CAN, LIN, FlexRay, and automotive Ethernet are natively supported, enabling robust testing of in-vehicle networks. Its modular hardware architecture makes it scalable for both component-level and full-system validation. SCALEXIO excels in ECU validation, fault injection experiments, and automated hardware certification workflows, making it a cornerstone platform for OEMs and suppliers focused on safety-critical autonomous systems.
- OPAL-RT RT-LAB: This platform specializes in high-fidelity real-time simulation leveraging multi-core CPUs and FPGA acceleration. RT-LAB is engineered for demanding applications that require ultra-low-latency and high-resolution timing, such as electric vehicle (EV) powertrain systems, motor control, energy storage management, and hybrid-electric systems. One of its key strengths is the ability to accurately model complex electrical and electro-mechanical systems with real-time feedback, enabling engineers to replicate in-vehicle transient behaviors like regenerative braking or inverter switching dynamics. RT-LAB offers deep integration with MATLAB/Simulink for model development and execution, and it supports a wide variety of real-time communication protocols including CAN, LIN, EtherCAT, and more. It interfaces smoothly with physical devices such as power amplifiers, load banks, and embedded ECUs, allowing it to serve as a central test platform for both component-level and full-system HIL validation in automotive, aerospace, and industrial automation domains.
- NI VeriStand: Developed by National Instruments, VeriStand is a comprehensive real-time testing environment designed for deploying and managing HIL systems. It enables engineers to execute real-time test sequences, interact with hardware I/O, and monitor system responses using a user-friendly interface. Built to work seamlessly with LabVIEW and TestStand, VeriStand allows for customizable test automation, signal conditioning, and calibration workflows. It is especially effective in environments utilizing NI PXI hardware, offering precision timing, deterministic execution, and support for a wide range of automotive communication protocols like CAN, LIN, and FlexRay. Engineers can also integrate FPGA-based components and model-in-the-loop capabilities, making VeriStand a versatile platform for validating embedded controllers, verifying signal integrity, and simulating real-world operating conditions under strict temporal constraints.
- Ansys AVxcelerate: While often recognized for its sophisticated sensor simulation capabilities, AVxcelerate also supports Hardware-in-the-Loop (HIL) applications through real-time emulation of sensor outputs such as LiDAR, radar, and camera feeds. These outputs can be interfaced with physical ECUs or perception stacks, enabling closed-loop HIL testing under a wide range of virtual driving scenarios including adverse weather, night driving, and dynamic traffic flows. The platform is designed for advanced sensor fusion validation, where multiple sensor modalities must be tested for coordination and redundancy. AVx-

celerate integrates with driving simulators and real-time testing environments to facilitate scenario-based safety validation, making it an ideal tool for verifying the behavior of safety-critical features like emergency braking, lane-keeping, and object avoidance in realistic yet controllable conditions. Its combination of photorealistic rendering and physical sensor behavior modeling ensures high-fidelity input for downstream HIL applications.

- AirSim[20]: An open-source platform developed by Microsoft and designed primarily for simulation of autonomous drones and ground vehicles using the Unreal Engine. While AirSim is not a native HIL platform, it provides extensive APIs in C++ and Python, along with support for the Robot Operating System (ROS), which enables the integration of custom hardware interfaces. Researchers can extend AirSim's simulated environments to include physical control systems, allowing them to evaluate navigation, control, and perception algorithms in limited HIL configurations. For example, an actual embedded controller can be connected to AirSim to send motor commands and receive simulated sensor feedback, emulating basic closed-loop behavior. This extensibility makes AirSim valuable for academic research, especially in early-stage prototyping where high visual realism and flexible simulation control are priorities.
- Project Chrono (HIL-capable extensions)[22]: An open-source physics engine designed for advanced multi-body dynamics simulation, Project Chrono provides highly detailed models of vehicle systems including tire-road interactions, suspension dynamics, and terrain effects. While not originally built as a HIL platform, it includes real-time execution modes and APIs that allow integration with external hardware through middleware like ROS, FMI, and real-time communication layers. Project Chrono supports real-time execution via synchronization with high-frequency solvers and latency-constrained threads, enabling hardware-in-the-loop interactions that depend on deterministic simulation updates. Middleware such as ROS bridges and Simulink co-simulation interfaces help Chrono exchange real-time data with ECUs or microcontrollers, supporting closed-loop testing workflows. When augmented with sensor emulation and real-time I/O hardware, Project Chrono can support academic-grade HIL experimentation. This makes it especially suitable for validating vehicle dynamics models, control algorithms, or robotics components where physical realism and multi-physics simulation are critical. Due to its extensibility and research-oriented architecture, it is frequently used in universities and labs exploring novel AV systems and human-in-the-loop driving studies.

Tools like IPG CarMaker [9] provide real-time testing environments for vehicle dynamics, sensor simulation, and ECU validation under diverse traffic scenarios.

12.3 Testing an Autonomous Vehicle

Autonomous vehicle systems require rigorous, specialized testing approaches that extend far beyond traditional unit and integration testing. Unlike conventional automotive systems, AVs must operate safely and intelligently in complex, open-ended environments with unpredictable human behavior and rapidly changing contexts. As such, effective testing must account for a wide range of uncertainties, including sensor imperfections, dynamic traffic conditions, infrastructure variability, and real-time computational constraints.

The critical challenge in AV testing lies in the fact that safety violations may stem from rare or combinatorial edge cases that are unlikely to be captured in routine testing. Therefore, AV testing must embrace a holistic strategy—combining realistic scenario-based assessments, granular subsystem validations, and statistically informed large-scale simulations. This section focuses on three essential techniques that reflect the distinctive demands of AV development: scenario-based testing for behavioral assessment, subsystem-specific validation for functional integrity, and accelerated or statistical testing for long-tail safety assurance and regulatory readiness.

12.3.1 Scenario-based Testing

Scenario-based testing involves the structured simulation and evaluation of AV behavior in predefined driving contexts, such as urban intersections, highway merging, pedestrian crossings, or sudden occlusions. These scenarios are central to understanding how AVs perform in specific, bounded interactions. Scenarios may be derived from naturalistic driving datasets, synthetically generated through modeling tools, or designed to meet regulatory benchmarks such as those from Euro NCAP or NHTSA. With the increasing complexity of AV systems, scenario formalization has become a critical discipline. Tools like SCENIC and frameworks such as ASAM OpenSCENARIO and OpenDRIVE are used to encode scenarios at different levels of abstraction—from logical to concrete—ensuring they are both human-interpretable and machine-executable. This standardization is crucial for ensuring repeatability, portability, and scalability across simulation engines, SIL/HIL setups, and safety case frameworks.

By focusing on how an AV performs in specific traffic or environmental conditions, scenario-based testing enables developers to evaluate the system's behavior under both common and exceptional situations. This technique provides a framework to test how an AV interprets, reacts, and adapts to stimuli within the environment, including the presence of pedestrians, other vehicles, cyclists, signage, and adverse weather conditions. Developers can examine whether the vehicle maintains safe margins, abides by road rules, and handles transitions between behaviors (e.g., from highway merging to lane changing) correctly.

These tests are especially crucial for validating decision-making logic in edge conditions—those rare but critical events that can provoke system-level failures. Scenario-based testing helps ensure consistent policy adherence and supports formal verification through coverage-based metrics. Edge case scenarios such as unexpected jaywalking, sensor noise or corruption, construction detours, and ambiguous intersections are carefully engineered to expose vulnerabilities. Such evaluation builds confidence that AVs will maintain safety, robustness, and reliability across the wide variability of real-world operation.

Recent state-of-the-art approaches have expanded scenario generation and evaluation methodologies:

- AdvSim[23] introduces adversarial perturbations in actor trajectories to uncover safety-critical weaknesses in the autonomy stack. It perturbs the motion paths of surrounding agents in traffic scenes to generate variations that are still plausible but are algorithmically designed to be more likely to induce failures in perception, prediction, and planning subsystems. This is accomplished using gradient-based or heuristic search optimization, which identifies those variations that expose blind spots in the AV's behavior models. The significance of AdvSim lies in its ability to systematically create edge-case scenarios without relying on hand-crafted or manually designed failure conditions. It addresses a key shortcoming in conventional scenario-based testing by prioritizing adversarial realism—crafting situations that are likely to occur but challenge AV safety boundaries. In doing so, it advances both the coverage and the efficiency of scenario-based safety validation, especially in multi-agent environments where interaction complexity often defies deterministic scripting.
- ChatScene[25] uses large language models to convert natural language traffic descriptions into testable code in simulators like CARLA. It addresses the challenge of low scenario diversity and manual scenario engineering by automating the translation of rich, high-level textual descriptions into structured simulation code. This enables rapid generation of a broad range of plausible traffic scenarios, including uncommon and imaginative edge cases. The significance of ChatScene lies in its ability to bridge human knowledge and simulator-executable format, allowing users to express testing intent in natural language while preserving fidelity in scenario structure. This approach is especially useful for enriching simulation datasets, stress-testing autonomy stacks with unusual conditions, and enabling scalable human-in-the-loop validation.
- ISS-Scenario[13] combines parameterized scenarios with random sampling and genetic algorithms to exhaustively evaluate behavior. It addresses the limitations of static scenario libraries and hand-crafted tests by automating large-scale behavioral validation in a structured yet diverse fashion. The framework enables users to define scenario templates with adjustable parameters, such as vehicle positions, speeds, and traffic densities. Through random sampling and evolutionary optimization, ISS-Scenario explores combinations that reveal performance limitations in planning, perception, or prediction modules. The significance of this approach lies in its ability to balance scenario generality with directed

search, systematically uncovering edge-case vulnerabilities while preserving plausibility. It is particularly effective in batch-testing pipelines where exhaustive and reproducible coverage is critical for regression analysis and continuous system validation.

- SCENIC[7] + VERIFAI[4] support probabilistic specification of scenarios and guided exploration for prediction model testing. SCENIC, a domain-specific probabilistic programming language, enables users to define scenarios with structured randomness—capturing both typical and edge-case behaviors in a formal and repeatable way. VERIFAI leverages these scenario specifications to systematically explore the space of inputs, using both random sampling and optimization techniques like falsification and counterexample generation. This approach solves the problem of limited scenario diversity in conventional testing and is especially significant for validating machine learning-based components, such as behavior prediction models, where performance is sensitive to subtle variations in input context. Together, SCENIC and VERIFAI offer a rigorous and scalable method to uncover hidden failure modes in high-dimensional, uncertain environments.
- STRIVE[17] leverages a learned traffic prior to generate plausible, accident-prone scenarios for AV planners. It addresses the challenge of constructing realistic but adversarial test cases by learning statistical patterns of traffic interactions from real-world data and then perturbing these in subtle but safety-critical ways. Unlike handcrafted or rule-based scenario generators, STRIVE produces scenarios that maintain ecological validity while pushing the limits of the AV planner's decision boundary. This is significant because it reveals weaknesses that only emerge under tightly coupled interactions between agents—such as multi-vehicle occlusion, ambiguous right-of-way decisions, or staggered lane changes. By focusing on plausible accident causality rather than random or extreme perturbations, STRIVE offers a principled, data-driven method to evaluate the robustness of planning algorithms in the scenarios most likely to lead to real-world failures.

These tools and frameworks emphasize not just coverage breadth but behavioral richness, increasing the likelihood of exposing vulnerabilities early in development cycles.

12.3.2 Accelerated and Statistical Testing

Given the infeasibility of physically testing AVs over billions of miles, accelerated testing methods simulate massive numbers of driving hours and scenarios in compressed time. These approaches aim to expose the AV system to rare, high-risk conditions more frequently and efficiently than naturalistic driving. Key techniques include:

- Monte Carlo Simulation: Running a scenario multiple times with randomized variables (e.g., pedestrian timing, weather conditions) to measure statistical performance and failure probability.
- Importance Sampling: Emphasizes rare but impactful events by biasing scenario generation to stress high-risk conditions. A foundational technique, Importance Sampling modifies distribution parameters to increase the frequency of rare events.
- Deep Importance Sampling: Builds upon Importance Sampling using neural networks to learn optimal sampling distributions in high-dimensional spaces, drastically improving efficiency and sample coverage.
- Closed-loop Falsification: Actively searches for failure conditions by modifying input parameters to 'falsify' system specifications, identifying previously unseen system vulnerabilities.
- Recent contributions by Huei Peng and others have further formalized accelerated testing strategies. These include:
- Accelerated Deployment[26]: A field-deployable framework for safely testing pre-production AVs on public roads by identifying and prioritizing deployment in environments that are statistically more likely to yield safety-critical interactions. Rather than uniformly testing across all geographic regions or road types, this method uses data-driven metrics to dynamically select locations where the AV is most likely to encounter complex traffic behaviors, rare events, or performance-limiting conditions. This targeted strategy addresses the inefficiency of conventional field testing by significantly increasing the density of informative interactions per mile driven. The approach is particularly significant because it bridges the gap between simulation-based acceleration and real-world validation, enabling manufacturers to gather actionable safety data with fewer miles and reduced risk exposure while supporting faster iteration cycles in the deployment pipeline.
- Scenario-Based Accelerated Testing: Combines scenario-specific testing with statistical acceleration, allowing for both targeted behavior evaluation and coverage of rare failures. This approach addresses the inefficiency of conventional simulation testing by leveraging structured scenario templates and stochastic variability to expose safety-critical system behaviors. Unlike generic Monte Carlo methods that sample from broad distributions, scenario-based acceleration prioritizes meaningful variation within high-risk scene categories—such as near-miss merges, occluded pedestrian crossings, or dense traffic maneuvers. This increases test relevance while retaining statistical robustness. The significance of this approach lies in its ability to couple human-relevant narratives (e.g., driver intent or social norms) with machine-driven diversity and rigor, producing tests that are both interpretable and exhaustive. It serves as a bridge between deterministic behavioral evaluation and scalable risk quantification, facilitating more actionable insights into AV performance under realistic and safety-critical constraints.

Recommended Papers to Read

1. Kaur, P., Taghavi, S., Tian, Z., & Shi, W. (2021, April). A survey on simulators for testing self-driving cars. In *2021 fourth international conference on connected and autonomous driving (MetroCAD)* (pp. 62–70). IEEE. [10]
2. Lou, G., Deng, Y., Zheng, X., Zhang, M., & Zhang, T. (2022, November). Testing of autonomous driving systems: where are we and where should we go? In *Proceedings of the 30th ACM joint European software engineering conference and symposium on the foundations of software engineering* (pp. 31–43). [15]
3. Hu, X., Li, S., Huang, T., Tang, B., Huai, R., & Chen, L. (2023). How simulation helps autonomous driving: A survey of sim2real, digital twins, and parallel intelligence. *IEEE Transactions on Intelligent Vehicles, 9*(1), 593–612. [8]
4. Li, Y., Yuan, W., Zhang, S., Yan, W., Shen, Q., Wang, C., & Yang, M. (2024). Choose your simulator wisely: A review on open-source simulators for autonomous driving. *IEEE Transactions on Intelligent Vehicles*. [14]

References

1. Amini, A., et al. (2022). Vista 2.0: An open, data-driven simulator for multimodal sensing and policy learning for autonomous vehicles. In *2022 international conference on robotics and automation (ICRA)* (pp. 2419–2426). IEEE.
2. AVL SiL. (2025). https://www.avl.com/en-us/testing-solutions/all-testing-products-and-software/connected-development-software-tools/avl-sil-suite.
3. Dosovitskiy, A., et al. (2017). CARLA: An open urban driving simulator. In *Conference on robot learning PMLR, 2017* (pp. 1–16).
4. Dreossi, T., et al. (2019). Verifai: A toolkit for the formal design and analysis of artificial intelligence-based systems. In *International conference on computer aided verification* (pp. 432–442). Springer.
5. dSPACE VEOS. (2025). https://www.dspace.com/en/inc/home/products/sw/simulation_software/veos.cfm.
6. ETAS COSYM. (2025). https://www.etas.com/ww/en/products-services/software-development-tools/cosym/.
7. Fremont, D. J., et al. (2019). Scenic: A language for scenario specification and scene generation. In *Proceedings of the 40th ACM SIGPLAN conference on programming language design and implementation* (pp. 63–78).
8. Hu, X., et al. (2023). How simulation helps autonomous driving: A survey of sim2real, digital twins, and parallel intelligence. *IEEE Transactions on Intelligent Vehicles, 9*(1), 593–612.
9. IPG Automotive. (2025). https://www.ipg-automotive.com/en/products-solutions/software/carmaker/.
10. Kaur, P., et al. (2021). A survey on simulators for testing self-driving cars. In *2021 fourth international conference on connected and autonomous driving (MetroCAD)* (pp. 62–70). IEEE.
11. Koenig, N., & Howard, A. (2004). Design and use paradigms for gazebo, an open-source multi-robot simulator. In *2004 IEEE/RSJ international conference on intelligent robots and systems (IROS) (IEEE Cat. No. 04CH37566)* (Vol. 3, pp. 2149–2154). IEEE.
12. Krajzewicz, D., et al. (2002). SUMO (Simulation of Urban MObility)-an open-source traffic simulation. In *Proceedings of the 4th middle East symposium on simulation and modelling (MESM20002)* (pp. 183–187).

13. Li, R., Qin, T., & Widdershoven, C. (2024). ISS-Scenario: Scenario-based testing in CARLA. In *International symposium on theoretical aspects of software engineering* (pp. 279–286). Springer.
14. Li, Y., et al. (2024) Choose your simulator wisely: A review on open-source simulators for autonomous driving. *IEEE Transactions on Intelligent Vehicles, 9*, 4861–4876.
15. Lou, G., et al. (2022). Testing of autonomous driving systems: Where are we and where should we go? In *Proceedings of the 30th ACM joint European software engineering conference and symposium on the foundations of software engineering* (pp. 31–43).
16. MATLAB Simulink. (2025). https://www.mathworks.com/products/simulink.html.
17. Paatil, P., et al. (2024). STRIVE: A co-simulation-based testing platform enhanced with runtime monitors. In *2024 16th international conference on COMmunication systems and NETworkS (COMSNETS)* (pp. 1106–1111). IEEE.
18. Rong, G., et al. (2020). Lgsvl simulator: A high fidelity simulator for autonomous driving. In *2020 IEEE 23rd international conference on intelligent transportation systems (ITSC)* (pp. 1–6). IEEE.
19. SCANeR Studio. (2025). https://www.avsimulation.com/en/scaner/.
20. Shah, S., et al. (2018). Airsim: High-fidelity visual and physical simulation for autonomous vehicles. In *Field and service robotics: Results of the 11th international conference* (pp. 621–635). Springer.
21. Siemens Prescan. (2025). https://plm.sw.siemens.com/en-US/simcenter/autonomous-vehicle-solutions/prescan/.
22. Tasora, A., et al. (2016). Chrono: An open source multi-physics dynamics engine. In *High performance computing in science and engineering: Second international conference, HPCSE 2015, Soláň, Czech Republic, May 25–28, 2015, Revised Selected Papers 2* (pp. 19–49). Springer.
23. Wang, J., et al. (2021). Advsim: Generating safety-critical scenarios for self-driving vehicles. In *Proceedings of the IEEE/CVF conference on computer vision and pattern recognition* (pp. 9909–9918).
24. Xu, R., et al. (2021). Opencda: An open cooperative driving automation framework integrated with co-simulation. In *2021 IEEE international intelligent transportation systems conference (ITSC)* (pp. 1155–1162). IEEE.
25. Zhang, J., Xu, C., & Li, B. (2017). Chatscene: Knowledge-enabled safety-critical scenario generation for autonomous vehicles. In *Proceedings of the IEEE/CVF conference on computer vision and pattern recognition* (pp. 15459–15469).
26. Zhao, D., et al. (2017). Accelerated evaluation of automated vehicles in car-following maneuvers. *IEEE Transactions on Intelligent Transportation Systems, 19*(3), 733–744.

Chapter 13
Industry Landscape

13.1 The State of the Autonomous Vehicle Industry

The automotive industry is undergoing one of the most significant technological shifts in history, marked by the convergence of advanced computing, artificial intelligence, and real-time data processing. The development of autonomous vehicles (AVs) has evolved from a speculative vision to a vibrant sector attracting significant investment and interdisciplinary collaboration. Today, the AV industry includes stakeholders ranging from global technology giants and established automobile manufacturers to innovative startups and governmental agencies. Their collective efforts are driven by the transformative potential of AVs: improved road safety, reduced traffic congestion, decreased emissions through optimized driving, and enhanced mobility for populations historically underserved by conventional transportation systems, such as the elderly or disabled.

Yet, the path to full autonomy remains fraught with complexity. Despite over a decade of research and billions of dollars in funding, no commercially available vehicle has achieved full autonomy (SAE Level 5). Most AVs operate in limited contexts—under specific conditions or within geofenced areas—relying on a blend of human oversight and pre-mapped environments. The industry's slow progression underscores the magnitude of the technological and systemic challenges it faces. These include the refinement of machine perception and decision-making in dynamic and unpredictable environments, the creation of robust safety validation frameworks, and the establishment of harmonized regulations across jurisdictions.

This chapter explores the current state of the AV industry with a focus on the key companies at its forefront, analyzing their respective technologies, market strategies, and deployment models. It will also address the regulatory, economic, and infrastructural hurdles that continue to shape the industry's trajectory. By doing so, the chapter provides a comprehensive overview of the ecosystem that defines the autonomous vehicle landscape today.

W. Shi, Y. He, *Introduction to Autonomous Driving*,
https://doi.org/10.1007/978-3-031-99485-2_13

13.2 Key Industry Players

The autonomous vehicle sector encompasses a dynamic and diverse array of stakeholders, each contributing distinct technological capabilities, resources, and strategic visions to the industry. These include major technology corporations with expertise in artificial intelligence and cloud infrastructure, traditional automotive manufacturers with deep experience in vehicle production and supply chain logistics, and a growing number of startups dedicated exclusively to developing autonomous driving systems. Some companies, like Tesla, integrate AV features directly into their consumer vehicles through progressive over-the-air updates, while others, such as Waymo and Zoox, aim to deploy fully autonomous robotaxis within controlled environments. Additionally, firms like Baidu represent a hybrid model, providing AV platforms and software to multiple automotive partners. This multi-pronged ecosystem reflects a broad spectrum of deployment philosophies—from enhancing existing human-driven vehicles with driver-assistance systems to leapfrogging directly into full autonomy for shared mobility services. The variety of approaches underscores the experimental nature of the industry and the absence of a universally accepted roadmap to autonomy.

13.2.1 Tesla: The Vision-Based Approach

Tesla has taken an aggressive and unique approach to autonomy, focusing on a vision-only system driven by cameras and neural networks, instead of relying on LiDAR and radar technologies like many of its competitors. This vision-based strategy, championed by CEO Elon Musk, posits that human-level autonomy can be achieved using only camera inputs processed through deep learning models. Tesla's Autopilot and Full Self-Driving (FSD) suite leverage vast amounts of real-world driving data collected from its global fleet, which is continuously fed into its neural networks to refine perception and decision-making algorithms. These models are updated and improved through over-the-air (OTA) software updates, enabling rapid iteration and deployment of new features.

This approach offers several advantages. The use of cameras makes the hardware setup more affordable and scalable, potentially allowing faster global deployment. Additionally, the real-time data collected from millions of vehicles helps Tesla train its AI models in diverse and unpredictable driving scenarios, fostering robust learning.

However, the strategy also has significant drawbacks. Critics argue that the absence of complementary sensors like LiDAR or radar could impair system reliability, especially in poor lighting or adverse weather conditions where cameras alone may struggle. The system's reliance on real-world data also means that edge cases—rare but critical driving scenarios—may not be adequately addressed until they are encountered in the wild. Regulatory scrutiny has intensified as Tesla markets its FSD package as nearly autonomous, even though it still requires active human supervision, raising safety concerns and debates over ethical transparency.

However, Tesla's strategy has been met with skepticism. The company markets FSD as a near-autonomous solution, yet its vehicles still require human supervision. Regulatory agencies have scrutinized Tesla's approach, and safety concerns persist regarding its reliance on purely vision-based AI systems.

13.2.2 Waymo: The Pioneer of Full Autonomy

Waymo, a subsidiary of Alphabet (Google's parent company), has established itself as a leader in AV technology, often viewed as a gold standard in the industry due to its cautious and data-driven approach. Unlike Tesla, which relies solely on cameras, Waymo employs a comprehensive sensor fusion strategy that combines high-definition LiDAR, radar, and cameras. This multimodal sensor suite provides redundant and complementary data streams, allowing the vehicle to perceive its environment with high precision and robustness, even in low-light or inclement weather conditions.

Waymo focuses primarily on developing robotaxis that are intended to operate without any human intervention. Its autonomous vehicles have been extensively tested in urban environments like Phoenix and San Francisco, where the company has launched limited commercial ride-hailing services. These environments are meticulously mapped and monitored to ensure system reliability.

The advantages of Waymo's approach include enhanced safety and perception accuracy due to sensor redundancy and rigorous validation protocols. The company also benefits from maintaining tight operational constraints—such as geofencing—to minimize unpredictable variables and optimize AV performance.

However, these strengths come with limitations. The high cost and complexity of the sensor hardware make scaling production more expensive. Moreover, the requirement for pre-mapped, controlled environments restricts deployment to a narrow set of locations, hindering rapid geographical expansion. As a result, while Waymo's AVs may be among the safest and most technically advanced, their operational flexibility and scalability remain constrained.

Waymo's vehicles are among the most advanced AVs in operation, but their deployment is limited to highly mapped geofenced areas. Scaling beyond these controlled zones presents significant challenges, particularly when it comes to adapting to unfamiliar and unpredictable environments.

13.2.3 Baidu: China's Autonomous Driving Giant

Baidu has emerged as a major player in the AV industry through its Apollo project, a comprehensive open-source platform designed to enable vehicle manufacturers to integrate autonomous driving technology seamlessly. Unlike many Western competitors who emphasize proprietary systems, Baidu has adopted a collaborative model that allows other companies to build upon its research and software infrastructure. This approach positions Baidu not just as a vehicle manufacturer but as a key enabler of AV adoption across the Chinese automotive landscape.

A major factor behind Baidu's rapid progress is the robust support it receives from the Chinese government, which provides favorable regulatory environments, access to urban infrastructure for testing, and large-scale pilot programs in smart cities. Baidu has already launched commercial robotaxi services in multiple Chinese cities, including Beijing and Wuhan, offering rides to the public in controlled zones.

This model presents several strengths. Government backing significantly reduces regulatory friction and speeds up deployment. The open-source nature of Apollo encourages innovation and standardization, while the company's partnerships with various automakers enable broad integration of AV features into a wide range of vehicles.

Nonetheless, Baidu's strategy has limitations. Its success is heavily reliant on national policy and localized support, which may not easily translate to international markets with different legal frameworks and consumer expectations. The collaborative platform, while open, could also introduce quality control challenges as third-party partners implement and modify Baidu's base technology. Moreover, while the robotaxi services have been well-publicized, they remain largely confined to geofenced areas with pre-established infrastructure, mirroring the scalability concerns faced by global counterparts.

Baidu's strategy emphasizes collaboration with automobile manufacturers, aiming to accelerate the adoption of AV technology across the industry rather than focusing solely on proprietary vehicles.

13.2.4 Zoox

Unlike other companies adapting existing vehicle designs, Zoox is developing a fully autonomous, bi-directional vehicle tailored specifically for ride-sharing. This vehicle is purpose-built from the ground up, featuring a symmetrical design with no traditional front or back, allowing it to move equally well in both directions. The cabin is designed for passenger comfort and efficiency, with no steering wheel or driver's seat, embodying the vision of a truly driverless experience.

The advantages of Zoox's approach lie in its architectural innovation and long-term scalability. By eliminating human-centric design constraints, the company can optimize space utilization, safety features, and sensor placement. Furthermore, the bespoke design enhances integration with fleet-based ride-hailing models, potentially reducing per-mile operational costs in the long run.

However, this strategy also presents considerable challenges. Developing a vehicle entirely from scratch requires extensive regulatory approvals and rigorous testing, often leading to long development timelines. Manufacturing such vehicles at scale is capital-intensive, especially when lacking the production infrastructure of legacy automakers. Additionally, public and regulatory acceptance of unconventional vehicle forms remains uncertain. While Amazon's acquisition of Zoox provides financial backing, the company must still prove that its novel approach can be commercially viable in a competitive and heavily scrutinized industry.

13.2.5 Other Key Players

- **Cruise (GM):** Backed by General Motors, Cruise focuses on electric AVs designed for urban environments. It had made notable strides by deploying self-driving taxis in San Francisco, operating in complex city traffic and accumulating operational data. However, in 2023, Cruise faced a significant setback when one of its autonomous vehicles was involved in a serious incident involving a pedestrian. This led to increased scrutiny from regulators and the public, prompting California authorities to suspend its driverless permits. Investigations revealed that the company's incident reporting had gaps and lacked sufficient transparency. Additionally, critics pointed to Cruise's aggressive deployment timeline and insufficient safety validation as contributing factors. The fallout forced Cruise to pause all autonomous operations nationwide, reassess its safety practices, and restructure parts of its leadership team. This episode underscored the difficulty of balancing rapid innovation with public safety and regulatory compliance in the AV space.
- **Argo AI (formerly Ford and VW-backed):** Argo AI was once considered a strong competitor but shut down in 2022 due to a convergence of strategic, financial, and technological challenges. Despite significant backing from Ford and Volkswagen and early technical successes in autonomous navigation and perception, the company struggled to commercialize its technology within a feasible timeframe. Investors grew increasingly concerned over the lack of a clear path to profitability and the regulatory uncertainties surrounding large-scale deployment. Moreover, the intense competition and high operational costs in the AV sector made sustained investment difficult to justify. Ultimately, Ford and Volkswagen decided to reallocate resources to in-house autonomous projects, leading to Argo AI's dissolution. The shutdown served as a stark reminder of the complexity of scaling AV technology from prototype to product.

13.3 Comparative Analysis of AV Deployment Strategies

Each AV company follows a different strategy for bringing autonomous technology to the market, shaped by their technological philosophies, resource availability, and long-term business goals. These strategies span a wide spectrum—from incremental improvements in advanced driver-assistance systems (ADAS) that gradually transition toward higher autonomy, to radical initiatives focused on launching fully autonomous ride-hailing services without steering wheels or human intervention.

Incremental strategies, such as those adopted by Tesla, allow companies to quickly deploy semi-autonomous features to consumer vehicles, gather extensive real-world driving data, and refine algorithms in the field. This data-driven approach supports rapid iteration and scalability but is also vulnerable to criticism for deploying partially autonomous systems in uncontrolled environments without adequate oversight.

On the other hand, companies like Waymo and Zoox pursue a top-down model where full autonomy is the end goal from the outset. These strategies often involve purpose-built vehicles, sensor fusion, and operation within geofenced areas to ensure safety and predictability. While these approaches prioritize safety and technical robustness, they suffer from high costs, regulatory hurdles, and limited geographical scalability.

Ultimately, the diverse deployment strategies reflect not only varying interpretations of technical feasibility but also differing perspectives on how best to gain public trust, regulatory approval, and economic viability in a rapidly evolving industry.

13.3.1 Tesla's Consumer-First Approach

Tesla aims to introduce autonomy incrementally by deploying semi-autonomous features in consumer vehicles, enabling it to roll out enhancements without waiting for fully validated Level 5 autonomy. This phased approach allows Tesla to utilize its extensive customer base as a massive real-world testing network. With millions of cars on the road, the company collects unprecedented volumes of real-world driving data, which are crucial for training and refining its neural network models. This iterative cycle of data collection and software updates via over-the-air (OTA) mechanisms provides Tesla with a flexible and fast development loop.

The advantages of this approach include rapid deployment, cost efficiency, and scalability. By incrementally rolling out features, Tesla can improve system performance and public familiarity with AV technologies while monetizing autonomous features through software upgrades.

However, this strategy also has notable drawbacks. The absence of full autonomy means that driver vigilance is still required, which can lead to overreliance and misunderstanding among users about the system's true capabilities. Critics argue that placing unfinished autonomous features in public hands introduces safety risks and creates a gray area in regulatory oversight. Additionally, Tesla's reliance on vision-only systems raises concerns about performance in edge cases, adverse weather, and low-visibility conditions where redundancy from other sensors could be beneficial. The regulatory landscape remains complex, with ongoing scrutiny from safety agencies about the marketing and performance of Tesla's Full Self-Driving (FSD) suite.

13.3.2 Waymo's Robotaxi Model

Waymo's strategy focuses on deploying fully autonomous ride-hailing services in highly controlled and geofenced environments. These urban testbeds—like Phoenix and San Francisco—are meticulously mapped and continuously monitored to ensure optimal performance of the AV systems. By limiting operations to predefined zones with high-definition maps and known variables, Waymo can maximize safety, reliability, and regulatory compliance.

The primary advantage of this method is its high standard of safety. Redundant sensor systems and detailed environmental models significantly reduce the likelihood of accidents and edge-case failures. This approach also enables better control over legal liabilities and operational conditions, allowing for smoother regulatory approval.

However, these strengths come at a cost. The necessity for exhaustive mapping and validation in each new city severely hampers scalability and market expansion. The approach also requires ongoing maintenance of map fidelity and infrastructure cooperation, which can be logistically and financially burdensome. Furthermore, the operational model may not be flexible enough to handle sudden changes in urban environments or adapt to regions with less structured road conditions. As a result, while Waymo's vehicles are among the safest and most advanced, the company faces significant hurdles in achieving broad geographic deployment and long-term commercial sustainability.

13.3.3 Baidu's Government-Backed Expansion

Baidu benefits from strong regulatory support in China, allowing it to scale AV deployments more aggressively. Unlike Tesla and Waymo, Baidu collaborates with multiple automakers, positioning itself as a technology provider rather than solely an AV operator.

13.4 Challenges in Scaling and Regulatory Compliance

Despite the rapid advancements in AV technology, significant barriers remain before widespread adoption is possible.

Despite the rapid advancements in AV technology, significant barriers remain before widespread adoption is possible. These obstacles span regulatory frameworks, societal acceptance, technical feasibility, and financial sustainability.

- **Regulatory and Legal Uncertainty:** One of the primary roadblocks to AV deployment is the lack of standardized regulations across regions and countries. Without clear definitions for legal liability in accidents involving AVs, manufacturers and operators face legal ambiguity. This uncertainty deters investment and slows down commercial rollout. Jurisdictions also vary significantly in their willingness to allow AV testing or deployment, creating a fragmented legal landscape.

- **Safety and Public Trust:** High-profile accidents involving autonomous systems have made the public cautious. While AVs aim to reduce human error, any failures—especially those that appear preventable—draw disproportionate scrutiny and raise ethical questions. Gaining public trust requires consistent, transparent performance and robust communication about the limits of current systems.
- **Infrastructure and Mapping Challenges:** AVs rely on high-definition maps, which must be continuously updated to reflect changing road conditions. In dynamic environments—where construction, traffic patterns, and signage can change frequently—this requirement poses a significant logistical and financial challenge. Additionally, urban and rural areas vary greatly in infrastructure readiness, limiting where AVs can safely operate.
- **Economic Viability:** Developing, testing, and scaling autonomous vehicle systems demand enormous capital. Most AV companies have not yet achieved profitability and remain reliant on parent companies or external investors. The costs associated with sensor suites, compute power, and extensive validation cycles make it difficult to deliver a cost-effective commercial product. Until these costs are reduced, broad consumer adoption will remain elusive.

13.5 Challenges in Human-Centered Deployment

In addition to regulatory fragmentation and technical limitations, real-world AV deployments increasingly expose human-centered challenges that are just as critical to scalability and public trust. A notable case study comes from the deployment of autonomous shuttles designed to serve vulnerable populations—such as older adults and individuals with physical or cognitive disabilities. Ren et al. [1] provides field-tested insights into the socio-technical barriers AV systems encounter when tasked with serving communities that stand to benefit most from improved mobility.

One of the most pressing lessons from this deployment is the inadequacy of treating accessibility as an afterthought. The shuttle deployment required deep architectural support for inclusive features, including physical accommodations (e.g., ramps and seating layouts), multimodal user interfaces (audio, visual, tactile), and redundancy in signaling and prompts. In practical terms, these adaptations were not luxuries—they were prerequisites for successful operation. This reveals a gap in how many AV platforms are currently designed and tested: simulation-based development environments rarely capture the lived experience of users with disabilities or sensory limitations.

Trust and rider comfort also emerged as non-trivial system requirements. For instance, slow and predictable vehicle behavior was preferred over speed or efficiency. This has implications not just for motion planning algorithms but for the design of entire HMI pipelines. Spoken prompts, pause logic, and feedback cues were essential for building passenger confidence. These findings challenge the industry trend of evaluating AVs primarily by throughput, efficiency, or disengagement metrics.

The study also demonstrates the irreplaceable role of community engagement. Unlike typical pilot programs that are evaluated via internal metrics, this shuttle service was co-designed with the very users it aimed to serve. Feedback sessions and stakeholder workshops led to route changes, UI redesigns, and interaction protocol improvements. This participatory model—while resource-intensive—yielded practical benefits and avoided design missteps that technical evaluations alone would not have caught.

From an industry landscape perspective, these insights highlight a persistent blind spot. While many AV firms focus on sensor fusion, planning architectures, or economic models, few allocate equivalent attention to user adaptation, especially among high-need groups. Yet it is precisely these groups that provide the most stringent test of system readiness and ethical robustness.

Incorporating these deployment lessons into the larger AV roadmap requires a recalibration of priorities: accessibility and human interaction must be treated as core performance criteria, not post-deployment optimizations. If AVs are to deliver on their promise of expanding mobility, they must work first—and best—for those who have been least served by the transportation systems of the past.

13.6 The Future of the AV Industry

The AV industry is expected to evolve significantly in the coming years as breakthroughs in artificial intelligence, sensor technologies, and edge computing enable more reliable and context-aware autonomous systems. These technical advancements will be accompanied by gradual progress in regulatory harmonization, enabling clearer legal frameworks and more predictable pathways for commercial deployment. Infrastructure investments, such as smart traffic systems and connected roadways, will further support the operational integration of AVs.

One likely outcome is the emergence of hybrid autonomy models that combine automated features with varying levels of human oversight depending on the driving context. Such models can serve as transitional solutions that mitigate safety risks while acclimating the public to increasingly autonomous systems. These could include systems where AVs manage routine driving tasks but hand control back to humans in complex or unstructured environments.

The future of autonomous vehicles is also inseparable from the evolution of vehicle computing systems. As autonomy matures, the computational backbone of vehicles must accommodate increasing demands for performance, reliability, and scalability. Vehicle computing is moving from fixed-function embedded systems to heterogeneous, modular platforms composed of CPUs, GPUs, TPUs, and dedicated accelerators, all interconnected through high-bandwidth internal networks. These systems must execute diverse and concurrent workloads, such as 3D perception, real-time localization, motion planning, and high-frequency control loops, often under strict timing and safety constraints.

A defining trend in vehicle computing is the shift toward co-designed architectures that integrate computation, communication, and control into a unified system. This co-design allows for intelligent resource allocation and supports the tight coupling of vehicle behavior with situational context. In practical terms, this means that decisions about where to process sensor data, when to communicate with the cloud or edge nodes, and how to control the vehicle must be orchestrated holistically. Autonomous vehicles are increasingly viewed not just as isolated platforms but as interconnected cyber-physical agents participating in a distributed, collaborative environment. This requires computing platforms to support dynamic workloads that can span in-vehicle processing units, edge infrastructure, and cloud data centers.

Real-time performance is another foundational requirement. Autonomous vehicles must respond to dynamic environments within strict temporal bounds. For instance, pedestrian detection, emergency braking, or evasive maneuvers demand end-to-end latency guarantees on the order of milliseconds. To meet this requirement, vehicle computing platforms must include real-time scheduling mechanisms that are both deterministic and adaptive. These mechanisms need to manage compute resources in the presence of unpredictable sensor input rates, varying network conditions, and competing task priorities. Recent frameworks introduce latency-aware schedulers capable of dynamically adjusting execution order and data flow to prioritize safety-critical tasks while maintaining overall system throughput.

In addition to performance, software abstraction is key to managing the complexity and variability of vehicle platforms. As vehicles incorporate diverse hardware and operating environments, there is a pressing need for standardized interfaces that separate application logic from the specifics of the underlying hardware. These abstractions enable portability, allowing perception, planning, or actuation modules to be deployed across different vehicles or hardware generations with minimal modification. This also facilitates modular updates and component verification, which are essential for maintaining functional safety and compliance with evolving regulatory standards.

Another pivotal direction is the treatment of the vehicle as a mobile edge node. Beyond supporting internal autonomy, vehicle computing platforms are being designed to host external services, such as cooperative perception, distributed mapping, and smart city analytics. This expands the functional role of autonomous vehicles into edge infrastructure, where they participate in data sharing, workload offloading, and environmental sensing. Such mobility-aware computing platforms must be capable of managing transient connectivity, protecting data privacy, and dynamically reallocating workloads as vehicles move through different network domains.

Supporting these functions at scale requires new middleware frameworks that ensure interoperability, security, and isolation. Programming interfaces must support composability, enabling developers to build reusable and certifiable software modules. Safety-critical tasks must be shielded from non-critical functions, such as infotainment or data logging, through strict partitioning. Moreover, these systems must be resilient to faults and capable of graceful degradation under resource constraints or partial failures.

The advantages of this evolutionary approach include increased flexibility, more manageable risk exposure, and higher public acceptance. However, it also presents challenges, such as maintaining user engagement in semi-autonomous modes and ensuring smooth transitions between human and machine control. Nonetheless, as the ecosystem matures, autonomous driving is poised to shift from a niche innovation to a core element of modern transportation infrastructure, impacting everything from urban mobility and freight logistics to accessibility and environmental sustainability.

Reference

1. Zhong, R., et al. (2024). *Autonomous shuttle operation for vulnerable populations: Lessons and experiences*. arXiv: 2402.17593 [cs.RO]. https://arxiv.org/abs/2402.17593.

Chapter 14
Conclusion

14.1 Vehicle Computing Is the Future: A Paradigm for Integrated Autonomy

As autonomous vehicles evolve, a transformative vision is taking hold—one that recognizes the vehicle itself not merely as a transport mechanism or isolated computational endpoint, but as a full-fledged, mobile computing platform embedded in a broader cyber-physical ecosystem. This vision is captured by the emerging Vehicle Computing (VC) paradigm[1, 2], which redefines how vehicles engage with their environment, infrastructure, and other agents. In contrast to conventional vehicle architectures that prioritize centralized processing or cloud reliance, VC enables distributed, real-time intelligence. Vehicles, under this model, are seen as computationally self-sufficient while remaining in constant interaction with peer vehicles and edge infrastructure. This dual capacity—localized autonomy with coordinated collaboration—marks a foundational departure in system design. The VC paradigm situates the vehicle as an active node within a distributed system of sensors, processors, energy reservoirs, and data sources. This reconceptualization supports new forms of real-time computation, cooperative sensing, and distributed intelligence that transcend conventional autonomous driving architectures.

A clear contrast between the traditional and vehicle computing paradigms is illustrated in Fig. 14.1. In the traditional closed-loop design, data flows linearly between sensors, control units, and hardware, creating a vertically integrated and functionally siloed architecture. In contrast, the vehicle computing paradigm introduces a decoupled structure, where vehicle data is shared bi-directionally with external sensors and infrastructure, routed through a vehicle programming interface, and then exposed to multiple application layers—including ADAS, system applications, and third-party services. This modularity and openness allow for more flexible, scalable, and service-oriented vehicle intelligence. The shift enables not just internal optimization but broader system-level integration with smart cities and cloud ecosystems.

W. Shi, Y. He, *Introduction to Autonomous Driving*,
https://doi.org/10.1007/978-3-031-99485-2_14

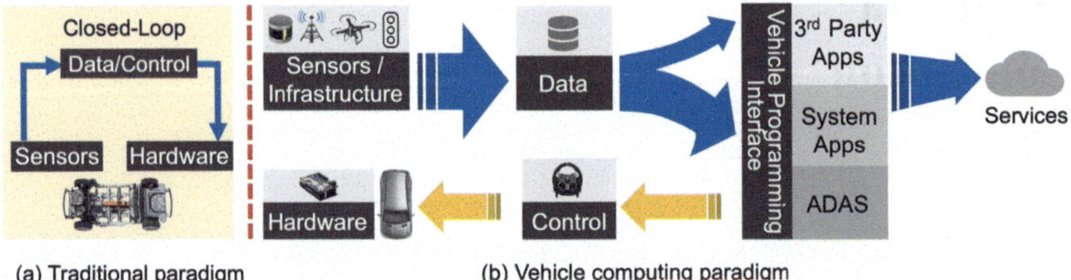

Fig. 14.1 Comparison of traditional vehicle control architecture (**a**) with the vehicle computing paradigm (**b**). The traditional model is based on a closed-loop system where sensors, data/control, and hardware form a tightly coupled, vertically integrated flow. In contrast, the vehicle computing paradigm enables modular, programmable access to vehicle functions through a vehicle programming interface, supporting data integration from external infrastructure and sensors, and enabling multi-layered services including ADAS, system apps, and third-party applications

At the heart of this paradigm lies the ACCESS principle: a fivefold framework encompassing computation, communication, energy, sensing, and storage. Each domain represents not just a technological function but a dynamic capability required for integrated autonomy. Computation in vehicle computing is no longer limited to traditional ECUs or basic control units—it includes high-throughput processors such as GPUs and NPUs that enable onboard inference for object detection, sensor fusion, and path planning. Communication under the VC model incorporates DSRC, C-V2X, and next-generation 5G/6G protocols to enable timely information exchange not just with infrastructure, but with pedestrians, mobile devices, and nearby vehicles. Energy management is broadened to include vehicle-to-grid (V2G) functionality, where EVs act as active participants in power distribution, capable of supplying energy during grid stress or disaster response. Sensing covers a suite of high-fidelity perception tools including LiDAR, millimeter-wave radar, and multi-spectrum cameras, providing data for environmental modeling and collaborative perception. Storage encompasses both low-latency buffers for real-time decision-making and high-capacity drives for longitudinal data collection—enabling diagnostics, safety forensics, and data-driven improvement of autonomy stacks.

Figure 14.2 visually encapsulates the ACCESS principle by presenting each of these five functional components as intrinsic to the vehicle's internal design. The image shows a conceptual vehicle with modules for computation (e.g., GPU, NPU, FPGA), sensing (LiDAR, camera, radar), communication, energy, and storage. This layout reinforces the notion that the modern connected vehicle integrates multiple subsystems into a tightly coupled computing platform. Each module supports real-time data generation, processing, and dissemination, providing the foundation for autonomous perception and decision-making.

Figure 14.3 scales this internal architecture outward to illustrate the broader Vehicle Computing ecosystem. It depicts how vehicles connect to other entities such as roadside units (RSUs), cellular towers, drones, edge servers, law enforcement infrastructure, and IoT devices. This diagram highlights the vehicle's role

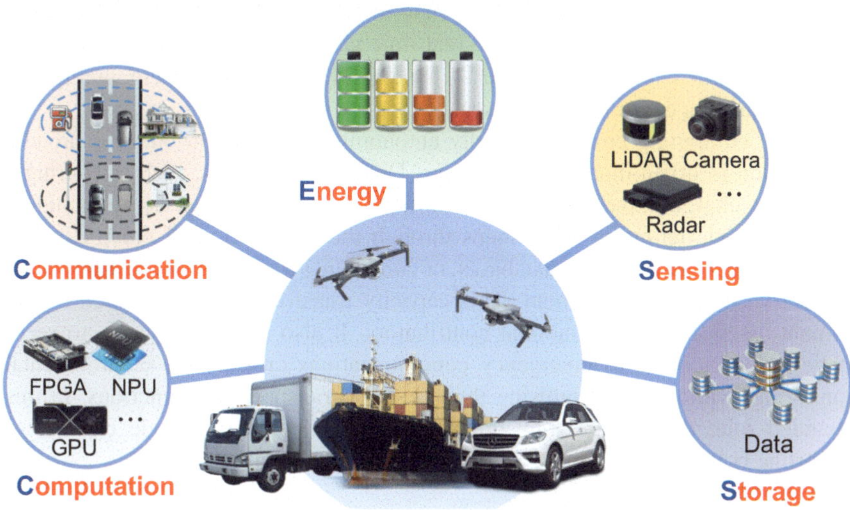

Fig. 14.2 Vehicle computing: Vehicle as a mobile computation, communication, energy consumption and storage, sensing, as well as data storage (ACCESS) platform

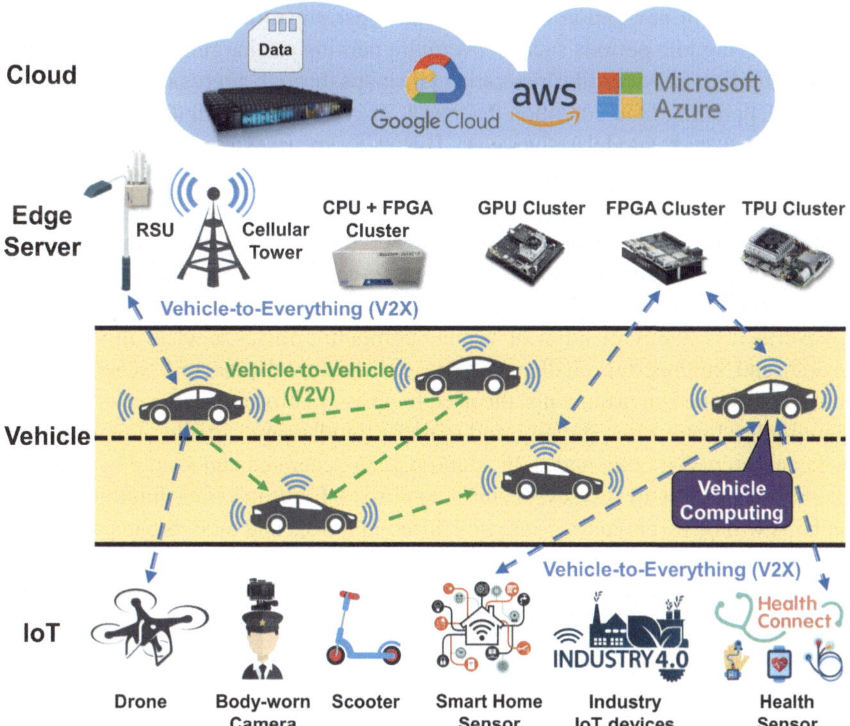

Fig. 14.3 The paradigm of vehicle computing

within a complex network of V2X (Vehicle-to-Everything) communication links. It illustrates data sharing flows—from real-time video and historical data exchange to environmental and safety messages—thus making clear how vehicle intelligence becomes part of a coordinated, city-scale cyber-physical system.

Vehicle computing enables not only autonomy in navigation but also broader participation in urban intelligence systems. During idle periods, such as overnight parking or charging, vehicles can be utilized for auxiliary services, including generating high-definition (HD) maps through the aggregation of previous route data, updating road condition databases, or conducting local machine learning tasks for fleet-wide model refinement. This capacity transforms parked vehicles from dormant assets into computational contributors. It also enables vehicles to serve public purposes, such as emergency communications or real-time environmental monitoring, without compromising their primary transportation functions. In this expanded role, the vehicle transitions into a civic computing node capable of supporting both private mobility and public data infrastructure.

14.2 Conclusion

The implications of this paradigm are far-reaching. Vehicle computing enables not only autonomy in navigation but also broader participation in urban intelligence systems. During idle periods such as overnight parking or charging, vehicles can be used for auxiliary services like generating HD maps through aggregation of previous route data, updating road condition databases, or conducting local machine learning tasks for fleet-wide model refinement. This capacity transforms parked vehicles from dormant assets into computational contributors. It also allows vehicles to serve public goals, such as emergency communications or real-time environmental monitoring, without detracting from primary transportation functions. In this expanded role, the vehicle transitions into a civic computing node capable of supporting both private mobility and public data infrastructure.

Nevertheless, the realization of vehicle computing raises a set of fundamental research and engineering challenges. These include real-time task scheduling in energy-constrained environments, the need for cross-platform runtime environments that unify hardware heterogeneity, and scalable middleware capable of maintaining low latency across mobile nodes. In addition, data generated and shared by vehicles must be managed with privacy, security, and auditability in mind. Interoperability standards will be essential to ensure that different makes and models can participate equitably in cooperative tasks. Moreover, the value chain around vehicle data will need to evolve, raising questions about ownership, compensation, and ethical constraints on usage. These challenges are not ancillary—they are central to realizing the full potential of the VC paradigm.

As autonomous vehicles (AVs) transition from experimental prototypes to operational systems in public spaces, the comprehensive body of knowledge presented in this book affirms one central insight: the development of autonomous driving is not solely a technical pursuit but an integrative, multidisciplinary endeavor. Chapter by chapter, we have seen how perception, planning, control, simulation, security, ethics, and systems engineering converge to support autonomy at scale. Each module, while individually complex, ultimately contributes to a system-level intelligence that must perform under diverse and uncertain real-world conditions.

Across the technical chapters, from sensor calibration and data fusion to real-time localization and trajectory planning, we have consistently emphasized the interplay between algorithmic sophistication and real-world constraints. Real-world autonomy is not simply a function of superior models or algorithms; it requires systems that can interpret noisy sensor inputs, accommodate partial observability, and operate under resource constraints. The integration of perception modules with predictive planning algorithms and fail-safe control mechanisms reflects the need for both depth and breadth in engineering design. For instance, the fusion of LiDAR, radar, and camera inputs is essential not only for redundancy but for a richer semantic understanding of complex environments.

The use of simulation frameworks like BlueICE has been highlighted throughout this text as essential for iterative development and validation. These platforms enable researchers and students to explore edge cases, compare algorithms under identical conditions, and simulate interactions with infrastructure components and human agents. The deterministic synchronization and co-simulation capabilities of such platforms bridge the gap between experimental reproducibility and deployment readiness. They also provide a safe environment to test rare or dangerous driving scenarios—conditions that are difficult to capture in real-world data.

Beyond technical achievement, the broader context explored in the chapters on societal impact, industry landscape, and V2X communication underscores the importance of cooperative intelligence. AVs do not operate in isolation; they exist in an ecology of human, digital, and infrastructural actors. The emergence of vehicle-to-everything (V2X) protocols, the integration of real-time edge computing systems, and the development of digital twins for infrastructure simulation are not auxiliary technologies but core enablers of scalable autonomy. These technologies shift the burden of intelligence from individual vehicles to distributed systems, highlighting a move toward cooperative autonomy where vehicles share intent, perception, and planned actions.

This book also addressed critical human-centered and ethical dimensions. Chapters on societal impact, security, and privacy emphasized the need to consider not just what AVs can do, but who they serve and how they affect broader communities. Ethical dilemmas—such as decision-making in unavoidable collision scenarios—cannot be resolved solely by algorithmic logic. They require public dialogue, regulatory oversight, and the inclusion of diverse perspectives. The exercises in ethical reflection are not peripheral but foundational to cultivating responsible autonomy.

Despite decades of progress, many open challenges remain. Full autonomy continues to be bounded by unresolved questions in interpretability, generalization, long-tail event handling, and ethical reasoning. Even with high-performing perception and control pipelines, rare events—such as erratic pedestrian behavior or sudden weather changes—can introduce unpredictable dynamics that challenge system robustness. Furthermore, questions of liability, transparency, and algorithmic fairness remain subjects of active debate and development. Ensuring that AVs operate equitably across geographies, demographics, and use cases will require continuous evaluation, public engagement, and policy innovation.

Yet, the technological and institutional momentum built through academic research, industrial deployment, and regulatory policy is unprecedented. The field is witnessing a shift from siloed, proprietary systems toward open interfaces, reproducible benchmarks, and collaborative standards. Initiatives such as open datasets, public competitions, and standardized evaluation protocols play a critical role in accelerating this transition. As a field, autonomous driving is moving beyond isolated prototypes and toward coordinated, accountable systems shaped by reproducible research and cross-sector collaboration.

Looking ahead, progress will depend on integrating emerging paradigms such as multimodal foundation models, open-world learning, secure-by-design architectures, and participatory design frameworks that include communities affected by AV deployment. The development of multimodal perception models—trained on vast datasets across vision, language, and sensor domains—offers promise for improving generalization and contextual reasoning. Similarly, open-world learning approaches aim to detect and adapt to novel conditions without exhaustive retraining, pushing the boundary of what AVs can autonomously recognize and handle.

Students, researchers, and engineers engaging with this book are encouraged to view autonomy not as a closed technical system but as an evolving sociotechnical infrastructure. The knowledge, tools, and principles laid out in these chapters aim to equip the next generation not only to innovate, but also to critically engage with the societal implications of mobility systems. This includes understanding historical inequities in transportation infrastructure, exploring participatory methods for deployment, and championing safety, inclusivity, and environmental sustainability.

Ultimately, autonomous driving challenges us to rethink not just how machines move through the world, but how we envision shared responsibility, infrastructure equity, and collective intelligence. The future of autonomy will be defined not only by technical capabilities, but by the inclusivity, resilience, and accountability of the systems we design. The next phase will demand interdisciplinary fluency, ethical commitment, and a collaborative spirit that spans engineering, urban planning, policy, and public engagement. This book provides a foundation—but the work ahead belongs to all of us.

References

1. Lu, S., & Shi, W. (2023). Vehicle as a mobile computing platform: Opportunities and challenges. *IEEE Network, 38*(6), 493–500.
2. Lu, S., & Shi, W. (2023). Vehicle computing: Vision and challenges. *Journal of Information and Intelligence, 1*(1), 23–35.

Index

W. Shi, Y. He, *Introduction to Autonomous Driving*,
https://doi.org/10.1007/978-3-031-99485-2

The manufacturer's authorised representative in the EU is Springer Nature Customer Service Centre GmbH, Europaplatz 3, 69115 Heidelberg, Germany. If you have any concerns regarding our products, please contact ProductSafety@springernature.com

Printed and bound by CPI Group (UK) Ltd, Croydon, CR0 4YY
07/07/2026
02160918-0001